LONDON MATHEMATICAL SOCIETY MONOGRAPHS
NEW SERIES

LONDON MATHEMATICAL SOCIETY MONOGRAPHS
NEW SERIES

Previous volumes of the LMS Monographs were published by Academic Press to whom all enquiries should be addressed. Volumes in the New Series will be published by Oxford University Press throughout the world.

NEW SERIES

1. *Diophantine inequalities* R. C. Baker
2. *The Schur multiplier* Gregory Karpilovsky
3. *Existentially closed groups* Graham Higman and Elizabeth Scott

Existentially Closed Groups

Graham Higman
Mathematical Institute,
University of Oxford

and

Elizabeth Scott
Department of Mathematics,
The Australian National University

CLARENDON PRESS · OXFORD
1988

Oxford University Press, Walton Street, Oxford OX2 6DP
Oxford New York Toronto
Delhi Bombay Calcutta Madras Karachi
Petaling Jaya Singapore Hong Kong Tokyo
Nairobi Dar es Salaam Cape Town
Melbourne Auckland
and associated companies in
Beirut Berlin Ibadan Nicosia

Oxford is a trade mark of Oxford University Press

Published in the United States
by Oxford University Press, New York

British Library Cataloguing in Publication data
Higman, Graham
Existentially closed groups.—(London Mathematical Society monographs. New series; no. 3).
1. Groups, Theory of
I. Title II. Scott, Elizabeth III. Series
512′.22 QA171
ISBN 0-19-853543-0

Library of Congress Cataloging in Publication Data
Higman. G. (Graham)
Existentially closed groups.
(London Mathematical Society monographs; new ser., no. 3)
Bibliography: p.
Includes index.
1. Existentially closed groups. I. Scott, Elizabeth L.
II. Title. III. Series
QA171.H62 1987 512′.2 87-7786
ISBN 0-19-853543-0

Typeset and printed in Northern Ireland by The Universities Press Belfast

Preface

This book grew out of a course of lectures given by Graham Higman during the academic year 1983–1984, the last year of his tenure of the Waynflete Professorship of Pure Mathematics at the University of Oxford. The course was given as an introduction to existentially closed groups, and was intended to be of interest to any mathematician with an interest in group theory or logic, while also being accessible to first-year graduate students. The book reflects this aim.

The theory of existentially closed groups is of interest in its own right; and it has an interesting history of changing relationships with the wider mathematical disciplines of which it forms a part. In its beginning, in the hands of B. H. Neumann and W. R. Scott, it formed part of the theory of infinite groups. Angus Macintyre, among others, made it a part of model theory; and M. Ziegler, in a fundamental work, transformed it into recursion theory. But some of the more recent work is very much in the spirit of the early years of the subject. Techniques from logic are used as necessary, but the primary aim is to find out the sorts of things that can go on in the infinite groups. This is true, for instance, of the important work of K. Hickin in this field; and also of the work of R. E. Phillips, and others, in the closely related and active field (not treated here) of relatively existentially closed groups. It is also the spirit of these notes, whose aim is not to present a systematic account of a finished subject; but to assist group theorists (research students or established mathematicians) to find their way into a corner of their subject which has its own characteristic flavour; and yet which is recognizably group theory.

We do not try to include everything that is known about existentially closed groups; we merely hope to give the reader sufficient understanding of, and interest in, the subject to inspire him to read further for himself. A list of original sources and further reading is given in the bibliography. We do not assume any prior knowledge of logic and all the recursive-function theory that will be needed is included in the text.

A large part of this book is based on the work of M. Ziegler, together with work by K. Hickin and A. Macintyre. But by no means all of their material is covered here. Large parts of Ziegler's (1980) paper remain beyond the scope of this book. For example, he gives a detailed classification of the $(\emptyset, \emptyset)$-generic groups. Hickin's (1985) paper did not appear until after this book had been written, and consequently none of the results of that paper are included here. Many other things have been omitted because a group-theoretical viewpoint has led us to concentrate on a different type of approach.

Some of the ideas in this book are, as far as we are aware, new; but most of them can be found, in some form or another, in earlier works. Generally, we do not give acknowledgments in the text for specific theorems, many of which are stated in different terminology, context, and generality from the original versions.

Existentially closed groups were first discussed by Scott (1951) and B. H. Neumann (1952 & 1973). Most of the basic results in the first chapter here are due to them.

The second chapter is based on the work of Hickin and Macintyre (1980), although some of the proofs are different.

Chapter 4 contains mostly basic results in recursion theory, which can be found in a textbook such as Rogers (1967).

The rest of the book is based broadly on the work of Ziegler (1980), although not all the results are his, (for example, Theorems 5.13 and 5.15 are due to Hickin and Macintyre (1980)). Many of the theorems can be found in Ziegler's papers in almost exactly the same form, but in some cases we have taken the ideas but treated them differently. In particular, Ziegler does not discuss games in general; all of his games are played under the finite code of rules. As a result he does not have R-generic groups. However, it should be noted that our $(\varnothing, \varnothing)$-generic existentially closed groups are what Ziegler, and others, call generic groups. Some of the results, for example The Relative-Subgroup Theorem (sometimes called The General Higman Embedding Theorem), were also proved by C. F. Miller, but Miller has never published this work.

These comments are intended only as a broad outline, to give the reader some idea of where to look for the original results. The history of many of these results is involved. For example, the theorem that a finitely generated group is embeddable in every existentially closed group if and only if it has solvable word problem was the combined result of work by Neumann (1973), Simmons (1973), and Macintyre (1972b); and some of the ideas in Chapter 9 originated with Macintyre (1972a). We can only apologize in advance for not undertaking the difficult task of always giving credit, even where it is undoubtably due.

Several people have been helpful during the writing of this book, but we would particularly like to thank Peter Neumann, whose timely interjections during the lectures ensured that proofs were corrected, and made more comprehensible. He also gave valuable advice during the writing of this book. We would also like to thank Laci Kovács, for the many helpful discussions that he and the second author had while the final draft of this book was being prepared in Canberra.

This book was begun in Oxford under funding from an SERC research grant, and it was completed while the second author held a fellowship at the Australian National University in Canberra. We would like to thank both of these institutions for their support.

Illinois G. H.
Canberra E. S.
October 1986

There are two particular debts that I must acknowledge here. One is to Wilfrid Hodges, who re-awakened my interest in this subject, when he lectured to the Aberdeen meeting of the British Mathematical Colloquium, in April 1983. Had I not found that talk stimulating, I would certainly not have lectured on this particular topic. The second debt, of course, is to my co-author, Dr E. A. Scott. As originally delivered, the lectures were very rough hewn. I left out chunks of necessary connecting material; the level of previous knowledge which I assumed of my audience varied widely and unsystematically; and I made a number of undeliberate mistakes. The task of reducing this mess to something publishable was immense and unenviable; but Dr Scott tackled it with energy and patience, and I am immensely grateful to her. (If I did not feel that what she has produced deserves well of the mathematical public, I would not continue the fiction of putting my name, with hers, on the title page of what is really her work.)

G. H.

Contents

Notation and conventions

We do not attempt to define the concepts discussed in this section, we assume that the reader is already familiar with them. We merely intend to establish the notation and conventions that we shall use in the text.

We denote the end of a proof with the symbol $\square$.

In general, maps will be written on the right and composed from left to right, so that xf is the image of x under f and $x(fg) = (xf)g$. In a few cases, primarily in Chapter 4, because of long standing precedents in logic and number theory, maps whose domains and ranges lie in the integers will be written on the left, and therefore composed from right to left, so that $(fg)(x) = f(g(x))$. In both cases the composition is informal in the usual sense that if $f: X \to Y$, and if $g: Y \to Z$ then $fg: X \to Z$ (if maps are being written on the right), or $gf: X \to Z$ (if maps are being written on the left) is always defined, we do not require that $f(X) = Y$.

We will use angle brackets to mean 'group generated by', so that $\langle g_1, \ldots, g_n \rangle$ denotes the group generated by elements $g_1, \ldots, g_n$ of some group. If R is a set of equations in $x_1, \ldots, x_n$ and their inverses, we write $\langle x_1, \ldots, x_n \mid R \rangle$ for the group generated by $x_1, \ldots, x_n$ subject to the defining relations R. We also write $\langle H_1, \ldots, H_n \rangle$ for the group generated by the subgroups $H_1, \ldots, H_n$. We may include groups on the left-hand side of the $\langle \bullet \mid \bullet \rangle$ notation, and in this case the relations of the group are always included in the defining relations, whether they specifically appear there or not. So that $\langle G, x_1, \ldots, x_n \mid R \rangle$ denotes the group generated by $x_1, \ldots, x_n$ and the elements of G, subject to the relations R and the relations of G.

We will often denote a sequence $(x_0, \ldots, x_n)$ or an infinite sequence $(x_0, x_1, x_2, \ldots)$ by $\boldsymbol{x}$, when it is clear from the context (or irrelevant) what the length and composition of the sequence is. Correspondingly, if w is a word in the letters $x_0, x_1, x_2, \ldots$ and their inverses, we will write $w(x_0, x_1, \ldots)$ or $w(x_0, \ldots, x_n)$ to indicate that the letters involved in w all lie in $\{x_0, x_1, \ldots\}$ or $\{x_0, \ldots, x_n\}$ respectively, and again we may often write $w(\boldsymbol{x})$ if the sequence concerned is clear from the context. We write $\mathrm{W}(x_0, x_1, \ldots)$ (or $\mathrm{W}(x_0, \ldots, x_n)$ or $W(\boldsymbol{x})$) for the set of all words on the x_i $(i \in \mathbb{N})$ (or $i = 0, \ldots, n$), and their inverses.

The following notation will be used:

(Here G, H, K are groups; A and B are Abelian groups; X, Y, X_i are sets; θ and ψ are group homomorphisms; and $g_0, \dots, g_n$ are elements of G)

$G \times H$	the direct product of G and H.
$G * K$	the free product of G and K.
$G \underset{H}{*} K$	the free product of G and K amalgamating the common subgroup H.
$A \oplus B$	the direct sum of A and B.
$C_G(g)$	$\{h \in G \mid h^{-1}gh = g\}$.
$C_G(H)$	$\{g \in G \mid g^{-1}hg = h;\quad \forall h \in H\}$.
$N_G(H)$	$\{g \in G \mid g^{-1}Hg = H\}$.
$H \leqslant G$	H is a subgroup of G.
$H \trianglelefteq G$	H is a normal subgroup of G.
G/H	the set of cosets, right or left as convenient, of H in G.
H^g	$\{g^{-1}hg \mid h \in H\}$.
$Z(G)$	$\{g \in G \mid g^{-1}hg = h;\quad \forall h \in G\}$.
Aut (G)	the automorphism group of G.
$\langle g_1, \dots, g_n \rangle^G$, $\langle H \rangle^G$	the normal subgroups of G generated by $g_1, \dots, g_n$ and $H \leqslant G$, respectively.
$H \cong G$	H is isomorphic to G.
G'	$\langle [g, h] \mid g, h \in G \rangle$.
$[g, h]$	$g^{-1}h^{-1}gh$.
ker θ, Im θ	the kernel and image of θ, respectively.
$\theta\|_H$	the restriction of θ to H.
$\theta \circ \psi$	the formal composition of θ and ψ.
$X \backslash Y$	$\{x \in X \mid x \notin Y\}$.
$\|X\|$	the cardinality, or order, of X.
$\bigcup_{i \in I} X_i$, $\bigcup X_i$	$\{x \mid \text{for some } i \in I,\quad x \in X_i\}$. The indexing set I may be omitted if it is clear from the context.
$\bigcap_{i \in I} X_i$, $\bigcap X_i$	$\{x \mid \text{for all } i \in I,\quad x \in X_i\}$.
$\varnothing$	the empty set.
$\mathbb{N}$	the natural numbers, $\{0, 1, 2, 3, \dots .\}$.
$\mathbb{Z}$	the integers, $\{\dots, -2, -1, 0, 1, 2, \dots .\}$.
$\mathbb{Q}$	the field of rational numbers.
$\mathbb{Q}^*$	$\mathbb{Q} \backslash \{0\}$, the multiplicative group of non-zero rationals.
$\mathbb{R}$	the real numbers.
$\mathbb{C}$	the complex numbers.
GL $(n, \mathbb{F})$	the group of all $n \times n$ invertible matrices of the field $\mathbb{F}$.

The following symbols are defined in the text; the page number refers to the page on which the symbol is first introduced.

	page		page
typ $(g_1, \ldots, g_n)$	7	$\langle G, \mathcal{S} \rangle$	58
Sk (G)	26	$\bar{P}$	87
$\mathcal{X}$	28	P^+, P^-	87
SC	28	$\Gamma(Q)$	95
JEP	28	$\mathrm{s}(X), \mathrm{s}(X; W)$	102
AEP	28	PW, PPW	106
AC	28	$\&$, *or*, $\neg$, $\exists$, $\forall$	110
$\mathcal{P}_{\mathrm{f}}(Z)$	35	$\mathrm{L}(a)$	111
$\leqslant_1, \equiv_1$	36	$\mathcal{C}, s\mathcal{C}$	114
$\mathbf{O}_{\mathrm{f}}(n)$	36	Alt $\mathbf{\Omega}$	115
$X \vee Y$	37	$\forall_n, \exists_n$	117
Rel $(a_1, \ldots a_r)$	38	Hom (x, y, u, v, s, t)	122
Rel (G)	39	$\mathbf{N}_{G,M}$	124
$\leqslant_{\mathrm{e}}, \equiv_{\mathrm{e}}$	40	$\mathfrak{A}$	126
M_J	44	ε_i	128
M_X	44	$U_i(X)$	128
$\leqslant_{\mathrm{T}}, \equiv_{\mathrm{T}}$	50	$\bar{\mathfrak{A}}$	129
$\leqslant^+, \equiv^+$	50	$St(x, y), St(x, y; X)$	129
G_X	51	$\mathfrak{A}^*$	135
$\leqslant^*, \equiv^*$	52	$\models, \Vdash, \Vdash_\infty$	141–143
$\mathrm{W}(x_1, x_2, \ldots), \mathrm{W}(\boldsymbol{x})$	53	$\mathrm{Th}_{\forall_n}(R), \mathrm{Th}_{\exists_n}(R)$	152
Ω	56		

A name followed by a date, e.g. (Smith, 1980) or Smith (1980), refers to a paper or a book which is listed in the bibliography.

1
Introduction

For a group G and variables $x, y, \ldots$, we use the expressions 'equation over G' and 'inequality† over G' in the natural sense. For example, if g and h are elements of G, then

$$x^3 = g$$

is an example of an equation over G, and

$$xgx^{-1} \neq h$$

is an example of an inequality over G. A set of equations and inequalities over G is said to be soluble in G if we can replace each variable by an element of G so as to make every member of the set true. The set is said to be soluble over G if it is soluble in some group containing G.

More formally, we can form the free product $G * F$, where F is the free group freely generated by the variables $x, y, \ldots$. Then an *equation* over G is a statement of the form $w = 1$, and an *inequality* over G is a statement of the form $w \neq 1$, where in each case $w \in G * F$. The elements of $G * F$ will be called *terms* over G. A set

$$\{w_i = 1, w_j \neq 1 \mid i \in I, j \in J\}$$

of equations and inequalities, over G, is *soluble over* G if there exists a group $H \supseteq G$ and a homomorphism

$$\theta : G * F \to H$$

such that

$$g\theta = g, \qquad w_j \notin \ker \theta, \qquad w_i \in \ker \theta,$$

for all $g \in G$, $i \in I$, and $j \in J$. The set is *soluble in* G if we can take $H = G$. Note that the above set is soluble over G if and only if the least normal subgroup of $G * F$ containing $\{w_i \mid i \in I\}$ intersects G trivially and contains no w_j, for $j \in J$.

† Some people use the term inequation here, but we prefer inequality, and so we will use it.

Examples

1. The set

$$\{x^{-1}gx = g, \quad x^{-1}hx = h, \quad x^{-1}kx \neq k\}$$

is not soluble over G if $g, h, k \in G$ and if $k = gh$.

2. If $C_G(g) \cap C_G(h) \subseteq C_G(k)$, but $k \notin \langle g, h \rangle$, then the above set is soluble over G but not in G.

3. The set $\{g = x^{-1}gx\}$ is soluble in G for any group G, and any element $g \in G$.

Definition 1.1 A group M is said to be *algebraically closed* if every finite set of equations defined over M that is soluble over M is soluble in M.

Definition 1.2 A group M is said to be *existentially closed* if every finite set of equations and inequalities defined over M that is soluble over M is soluble in M.

Let $\mathcal{B}$ be the propositional calculus generated by the propositions $w = 1$, where w is a term over G. By the Disjunctive Normal Form Theorem, anything in $\mathcal{B}$ is equivalent to a statement of the form

$$A_1 \text{ or } A_2 \text{ or } \cdots \text{ or } A_k,$$

where $A_j = p_{ij} \,\&\, \cdots \,\&\, p_{nj} \,\&\, q_{lj} \,\&\, \cdots \,\&\, q_{mj}$, in which p_{ij} is a statement $w_{ij} = 1$ and q_{lj} is a statement $v_{lj} \neq 1$ $(1 \leq i \leq n, 1 \leq l \leq m, 1 \leq j \leq k)$, with w_{ij} and v_{lj} terms over G. Thus Definition 1.2 is equivalent to the following.

Definition 1.2* A group M is said to be existentially closed if every finite formula in $\mathcal{B}$ that is satisfiable in some group containing M is satisfiable in M.

Our definition of an algebraically closed group differs from the definition which is often taken, in that we do not assume that $M \neq 1$. We have taken the above definition because, as we will show later in this chapter, the existentially closed groups are precisely the non-trivial algebraically closed groups.

There are three elementary embedding theorems which we will use extensively throughout these notes. We state the theorems here without proofs, but the proof of the second may be found in Higman, Neumann, and Neumann (1949) and the third in Britton (1963). In addition, all three theorems are proved, and discussed in detail, in Chapter IV of Lyndon and Schupp (1977).

A. Free products *If* $\{G_\alpha \mid \alpha \in T\}$ *is any set of groups, with* $G_\alpha \cap G_\beta = U$ *whenever* $\alpha \neq \beta$, *then there is a group containing* $\bigcup_{\alpha \in T} G_\alpha$. *In the 'freest' group containing this union, if* $g_1 \cdots g_n = 1$, *with* $g_j \in G_{\alpha_j}$, *then* $\alpha_i = \alpha_{i+1}$ *or* $g_i \in U$ *for some i.*

B. HNN-extensions *If* $\theta : A \to B$ *is an isomorphism between subgroups A and B of G then the group H generated by G and t subject to the relations*

$$t^{-1}at = a\theta \quad \text{for all } a \in A$$

contains G. The group H is called an HNN-extension of G.

C. Britton's Lemma *In the HNN-extension*

$$H = \langle G, t \mid t^{-1}at = a\theta; \forall a \in A \rangle$$

of G, if $g_0 t^{\varepsilon_1} g_1 \cdots t^{\varepsilon_n} g_n = 1$, *where* $\varepsilon_i = \pm 1$ *and* $g_j \in G$ $(1 \leqslant i \leqslant n, 0 \leqslant j \leqslant n)$, *then, for some* $q \in \{1, \ldots, n-1\}$, *either*

(*a*) $\varepsilon_q = -\varepsilon_{q+1} = -1$ *and* $g_q \in A$,

or

(*b*) $\varepsilon_q = -\varepsilon_{q+1} = 1$ *and* $g_q \in B$.

The rest of this chapter is devoted to proving some basic results about existentially closed groups. Our first aim is to prove that every non-trivial algebraically closed group is existentially closed.

Lemma 1.3 *If g is an element of an existentially closed group M, then there is a finite set of equations and inequalities that is soluble in M if and only if g has infinite order.*

Proof Consider the set $\mathscr{S} = \{x^{-1}gx = g^2,\ y^{-1}g^2y = g^2,\ y^{-1}gy \neq g\}$, with variables x and y. If $g \in M$ has infinite order, then $\langle g \rangle \cong \langle g^2 \rangle$; so we can form the HNN-extension $H = \langle M, x \mid x^{-1}gx = g^2 \rangle$. Now $g \notin \langle g^2 \rangle$ so we can form the HNN-extension $K = \langle H, y \mid y^{-1}g^2y = g^2 \rangle$, in which $y^{-1}gy \neq g$. Then $\mathscr{S}$ is soluble in K which contains M and so, by definition of M, $\mathscr{S}$ is soluble in M. If g has finite order and if $x^{-1}gx = g^2$ for some $x \in M$, then g has odd order. So then $g \in \langle g^2 \rangle$, and hence $y^{-1}g^2y = g^2$ implies that $y^{-1}gy = g$. Thus, either $x^{-1}gx = g^2$ or

$$\{y^{-1}g^2y = g^2,\ y^{-1}gy \neq g\}$$

is not soluble over M. Hence, $\mathscr{S}$ is soluble over M if and only if g has infinite order. □

Lemma 1.4 *Let M be an existentially closed group, and let each of m, n, p be either an integer greater than 1 or else infinite. Then M contains elements g, h, k, of orders m, n, p respectively, such that $gh = k$.*

Proof If m is finite, define $\mathcal{S}_m = \mathcal{S}_m(x_1, y_1, z_1)$ to be the set

$$\{x_1 \neq 1, \dots, x_1^{m-1} \neq 1, \quad x_1^m = 1\}.$$

If m is infinite, define $\mathcal{S}_m$ to be the set

$$\{z_1^{-1}x_1z_1 = x^2, \quad y_1^{-1}x_1^2y_1 = x_1^2, \quad y_1^{-1}x_1y_1 \neq x_1\}.$$

In either case we see that, if $\mathcal{S}_m$ is soluble in M, then x_1 has order m. Define $\mathcal{S}_n(x_2, y_2, z_2)$ and $\mathcal{S}_p(x_3, y_3, z_3)$ similarly, and let

$$\mathcal{S} = \mathcal{S}_m \cup \mathcal{S}_n \cup \mathcal{S}_p \cup \{x_1x_2x_3^{-1} = 1\}.$$

Let $A = \langle a, b, c\rangle$ be any group such that a, b, c have orders m, n, p respectively and $ab = c$. (If m, n, p are all finite, a proof that such a group A exists can be found in Coxeter and Moser (1972). The basic idea is to take a triangle (spherical or hyperbolic if necessary) whose angles are π/m, π/n, and π/p. Then let r, s, t be reflections in each of the sides of the triangle and take $a = rs$, $b = st$, $c = tr$. If m (say) is infinite, then we can take A to be $\mathbb{Z}_n * \mathbb{Z}_p$. For later use we also note that, if m is infinite, we could take $A = \langle s, t\rangle *_{\langle t\rangle} \langle t, r\rangle$, where $\langle s, t\rangle \cong D_{2n}$ and $\langle t, r\rangle \cong D_{2p}$.)

We can solve $\mathcal{S}$ in either $M * A$ (if m, n, p are all finite) or in some HNN-extension of $M * A$ (e.g. $\langle M * A, y_1, z_1\rangle$ if m is infinite). Thus, we can solve $\mathcal{S}$ in M, and if $\bar{x}_1, \bar{x}_2, \bar{x}_3$ are solutions, then we can take $g = \bar{x}_1$, $h = \bar{x}_2$, and $k = \bar{x}_3$. □

Lemma 1.5 *For any group G, and any elements g and h of G, the following statements are equivalent*:

(*i*) $(g = 1) \Rightarrow (h = 1)$,

(*ii*) *The equation $x^{-1}gxy^{-1}gy = h$ is soluble over G.*

(*Note*: *The implication in* (*i*) *is the logical implication, i.e. either $h = 1$ is true or $g = 1$ is false.*)

Proof Clearly (ii) implies (i). So suppose that $(g = 1) \Rightarrow (h = 1)$. Then either $h = 1$ or both $h \neq 1$ and $g \neq 1$.

If $h = 1$, then, given any $g \in G$, form the HNN-extension

$$K = \langle G, t \mid t^{-1}gt = g^{-1}\rangle.$$

Putting $y = 1$ and $x = t$, the equation $x^{-1}gxy^{-1}gy = h$ is soluble over G, in K.

If $h \neq 1$ and $g \neq 1$, suppose that g and h have orders m and n respectively, ($m > 1$ and $n > 1$ but m and n may be infinite). From Lemma 1.4, we see that there exists a group A generated by a, b, and c, with orders m, m, and n, respectively, such that $ab = c$. Since c and h

have the same order, we can form the free product

$$K = G \underset{h=c}{*} A$$

of G and A amalgamating $\langle c \rangle$ and $\langle h \rangle$. Then we can form the HNN-extensions

$$L = \langle K, t \mid t^{-1}gt = a \rangle \quad \text{and} \quad H = \langle L, s \mid s^{-1}gs = b \rangle.$$

Choosing $x = t$ and $y = s$, we see that the equation

$$x^{-1}gxy^{-1}gy = ab = c = h$$

is soluble in H, and hence over G. □

Lemma 1.6 *For elements $g_1, \ldots, g_k, h$, of a group G, the following two statements are equivalent:*

(*i*) $(g_1 = g_2 = \cdots = g_k = 1) \Rightarrow (h = 1)$

(*ii*) *The equation*

$$g_1^{x_1} g_1^{y_1} g_2^{x_2} g_2^{y_2} \cdots g_k^{x_k} g_k^{y_k} = h$$

has a solution over G.

Proof Again it is clear that (ii) implies (i). We show that it is a corollary of Lemma 1.5 that (i) implies (ii).

Suppose that $(g_1 = \cdots = g_k = 1) \Rightarrow (h = 1)$. If $h = 1$ then, by a sequence of k HNN-extensions of G, we can construct a group containing G in which the equation in (ii) is soluble. If $h \neq 1$ then $g_i \neq 1$, for some i, and for this i the statement

$$(g_i = 1) \Rightarrow (h = 1)$$

holds. Then, by Lemma 1.5, we can find a group G_1, containing G, in which the equation

$$x_i^{-1} g_i x_i y_i^{-1} g_i y_i = h$$

is soluble. There exists a repeated HNN-extension G_2 of G_1 in which the equation

$$g_1^{x_1} g_1^{y_1} \cdots g_{i-1}^{x_{i-1}} g_{i-1}^{y_{i-1}} = 1$$

is soluble, and there exists a repeated HNN-extension G_3 of G_2, in which the equation

$$g_{i+1}^{x_{i+1}} g_{i+1}^{y_{i+1}} \cdots g_k^{x_k} g_k^{y_k} = 1$$

is soluble. Thus the equation in (ii) is soluble in G_3, and hence over G. □

Theorem 1.7 *A group is existentially closed if and only if it is algebraically closed and non-trivial.*

Proof Clearly any existentially closed group is algebraically closed. The trivial group is not existentially closed, since the inequality $x \neq 1$ is soluble in some group, and therefore over, but not in, the trivial group.

Let M be a non-trivial algebraically closed group. Consider the equations and inequalities

$$\{u_1 = 1, u_2 = 1, \ldots, u_n = 1, v_1 \neq 1, v_2 \neq 1, \ldots, v_m \neq 1\}.$$

We suppose that this set is solvable over M, in K say. Because M is non-trivial, we can choose a non-trivial element h of M. The statement $v_i \neq 1$ is equivalent to the statement $(v_i = 1) \Rightarrow (h = 1)$, and, by Lemma 1.5, this is equivalent to the solubility of the equation $s_i^{-1}v_is_it_i^{-1}v_it_i = h$ over any group containing both v_i and h. (The elements s_i and t_i $(1 \leqslant i \leqslant m)$ are new variables not involved in the u_j or the v_i.) Thus, the set of equations $\{u_j = 1,\ \ h = s_i^{-1}v_is_it_i^{-1}v_it_i \mid 1 \leqslant j \leqslant n,\ \ 1 \leqslant i \leqslant m\}$ is soluble over K and hence over M. But M is algebraically closed, so this set has a solution in M and this solution necessarily satisfies $u_j = 1$ and $v_i \neq 1$, for $1 \leqslant j \leqslant n$ and $1 \leqslant i \leqslant m$. Thus M is existentially closed. □

It will be shown in Chapter 3 that existentially closed groups do exist. In fact, there are $2^{\aleph_0}$ countable existentially closed groups, any group can be embedded in an existentially closed group, and any countable group can be embedded in a countable existentially closed group.

Theorem 1.8 *Let M be an existentially closed group. Then we have:*

(*a*) *M cannot be finitely generated.*

(*b*) *M contains every finitely presented simple group (and hence every finite group).*

(*c*) *M is simple.*

Proof (a) Let $g_1, \ldots, g_k \in M$. We can solve the equations and inequalities $x^{-1}g_1x = g_1, \ldots, x^{-1}g_kx = g_k,\ \ x^{-1}yx \neq y$ over M (e.g. in the direct product $M \times G$, where G is any non-Abelian group) and hence in M. Then there exists some $y \in M$, such that $y \notin \langle g_1, \ldots, g_k \rangle = X$ and so $M \neq X$.

(b) Let $G = \langle g_1, \ldots, g_k \mid w_1(\mathbf{g}) = \cdots = w_r(\mathbf{g}) = 1 \rangle$ be a finitely presented group, containing a non-trivial element $u(\mathbf{g})$. Let $\{x_1, \ldots, x_k\}$ be a set of distinct variables. We can solve the set

$$\{w_1(\mathbf{x}) = w_2(\mathbf{x}) = \cdots = w_r(\mathbf{x}) = 1, \quad u(\mathbf{x}) \neq 1\}$$

over M (in $M \times G$), and hence in M. So M contains a non-trivial homomorphic image of G. If G is simple this must be G itself, and so G can be embedded in M.

(c) If $g,h \in M$ and $g \neq 1$, then $(g = 1) \Rightarrow (h = 1)$ holds in M. So, by Lemma 1.5, we can solve the equation $x^{-1}gxy^{-1}gy = h$ over M, and hence in M. Thus, every $h \in M$ lies in the normal subgroup generated by g, for any $g \neq 1$. So M is simple. □

Definition 1.9 If $\{g_1, \ldots, g_k\}$ and $\{h_1, \ldots, h_k\}$ are sets of elements which lie in groups G and H respectively, we write

$$\text{typ}\,(g_1, \ldots, g_k) = \text{typ}\,(h_1, \ldots, h_k),$$

and say that the groups $\langle g_1, \ldots, g_k\rangle$ and $\langle h_1, \ldots, h_k\rangle$ are of the *same type*, if there is an isomorphism from $\langle g_1, \ldots, g_k\rangle$ to $\langle h_1, \ldots, h_k\rangle$ which carries g_i to h_i for all $i \in \{1, \ldots, k\}$.

Theorem 1.10 *If M is an existentially closed group, and if $g_1, \ldots, g_k, h_1, \ldots, h_k$ are elements of M with* $\text{typ}\,(g_1, \ldots, g_k) = \text{typ}\,(h_1, \ldots, h_k)$, *then there exists an element t of M such that $t^{-1}g_it = h_i$ $(1 \leq i \leq k)$.*

Proof We can form an HNN-extension of M in which the set

$$\{x^{-1}g_ix = h_i \mid 1 \leq i \leq k\}$$

can be solved $\big(K = \langle M, t \mid t^{-1}g_it = h_i = g_i\theta \ (1 \leq i \leq k)\rangle\big)$. Thus, these equations can be solved in M. □

Definition 1.11 A group G is *SQ-universal* if every countable group is a subgroup of a quotient group of G.

The theorem (Higman *et al.*, 1949) that every countable group can be embedded in a 2-generator group, says precisely that the free group of rank 2 is SQ-universal. It is easy to see that a group which has an SQ-universal quotient group is SQ-universal. It is a result of P. M. Neumann (1973) that the triangle groups $\langle g, h \mid g^l = h^m = (gh)^n = 1\rangle$ are SQ-universal if and only if $(1/l) + (1/m) + (1/n) < 1$. We shall use this result in the proof of the next theorem.

A *local system* in a group G is a set of subgroups $\{H_i \mid i \in I\}$ such that:

(1). If $i,j \in I$ then there exists $k \in I$ with $\langle H_i, H_j\rangle \subseteq H_k$.

(2). The union of all the H_i is G.

If $\mathcal{P}$ is a property of groups, to say that G is *locally* $\mathcal{P}$ means that G has a local system, every member of which has the property $\mathcal{P}$. If G having $\mathcal{P}$ implies that G is finitely generated, then G is locally $\mathcal{P}$ if and only if every finitely generated subgroup of G is contained in a subgroup of G with $\mathcal{P}$. To see this, take $\{K_i \mid i \in I\}$ to be the set of all finitely generated subgroups of G. Then, if each K_i is contained in some H_i which has $\mathcal{P}$, and is therefore finitely generated, the set $\{H_i \mid i \in I\}$ is a local system in G, so G is locally $\mathscr{P}$. (The other way around, the result is trivial.)

We are now in a position to prove the last theorem of this chapter.

Theorem 1.12 *An existentially closed group is locally two-generator-and-perfect. (A group is perfect if it is equal to its own derived subgroup.)*

Proof Let $K = \langle g, h \mid g^2 = h^3 = (gh)^7 = 1 \rangle$, which is SQ-universal by the result of P. M. Neumann (1973), quoted above. If K' is the derived subgroup of K, then K/K' is Abelian and generated by $\bar{g} = gK'$ and $\bar{h} = hK'$. Now $\bar{g}^2 = \bar{h}^3 = (\bar{g}\bar{h})^7 = 1$, so $1 = \bar{g}^7\bar{h}^7 = \bar{g}\bar{h}$ and $\bar{g} = \bar{h}$. Thus $\bar{h}^2 = 1$ and $\bar{h} = \bar{g} = 1$. So $|K/K'| = 1$ and K, and hence each of its homomorphic images, is perfect. Let $G = \langle g_1, \dots, g_k \rangle$ be a finitely generated subgroup of M. Then G belongs to some homomorphic image $\langle a, b \rangle$ (say) of K, with $g_i = w_i(a, b)$. The equations $x^3 = y^2 = (xy)^7 = 1$ and $g_i = w_i(x, y)$ can thus be solved in the free product of M with $\langle a, b \rangle$, amalgamating G. So there exist $c, d \in M$ such that $G \subseteq \langle c, d \rangle \subseteq M$ and $\langle c, d \rangle$ is a homomorphic image of K. Thus, $\langle c, d \rangle$ is a 2-generator perfect subgroup of M containing G, and, by the remark above, M is locally 2-generator-and-perfect. □

2
Centralizers and normalizers of subgroups

This is a technical chapter in which we discuss some aspects of the subgroup structure of existentially closed groups. In particular, we contrast the natures of centralizers and normalizers of finite subgroups with those of infinite subgroups. All the results proved here are true for any existentially closed group.

First we aim to prove the following theorem.

Theorem 2.1 *If G is a finite characteristically simple subgroup of an existentially closed group M, then the normalizer of G in M is a maximal subgroup of M.*

The method that we will use to prove Theorem 2.1 is different to that originally given by Hickin and Macintyre (1980). In the original draft of this book we were only able to use this alternative method to prove the theorem in the case where G is a cyclic group of prime order. We are grateful to Ken Hickin for pointing out how these methods can be used to prove the full theorem.

Let M be any existentially closed group, and let G be any finitely generated subgroup of M. Let Γ be the set of all subgroups of M which are isomorphic to G. Then M acts by conjugation as a permutation group on Γ; the action must be faithful because M is simple. Since G is finitely generated, we see from Theorem 1.10 that, given $H \in \Gamma$, there exists an element $a \in M$ such that $H = G^a$. Thus, the action of M is transitive, and the group $\mathrm{N}_M(G)$ is a point stabilizer in this action. Since M acts transitively, $\mathrm{N}_M(G)$ is maximal if and only if M acts primitively, i.e. if and only if the only equivalence relations on Γ which are respected by the action of M are equality and the universal relation. So, letting $\equiv$ be any equivalence relation on Γ which is respected by M, we prove Theorem 2.1 by showing that if there exist G_1 and G_2 in Γ such that $G_1 \neq G_2$ but $G_1 \equiv G_2$, then $H_1 \equiv H_2$, for all H_1 and H_2 in Γ.

First we show that, if there exist G_1 and G_2 in Γ such that $G_1 \equiv G_2$ but $G_1 \neq G_2$, then we may assume that $G_1 \cap G_2 = 1$.

Lemma 2.2 *Let $G=\{g_1, \dots, g_n\}$ and $H=\{h_1, \dots, h_n\}$ be isomorphic finite characteristically simple subgroups of an existentially closed group M, such that $1<|G\cap H|<|G|$. Then we can choose a subgroup K and an element a of M such that $G^a=K$, $H^a=H$ and $|G\cap K|<|G\cap H|$.*

Proof We may suppose that $\operatorname{typ}(g_1, \dots, g_n)=\operatorname{typ}(h_1, \dots, h_n)$. Since $G\cap H$ is a non-trivial subgroup of H we can choose an automorphism of H which does not fix $G\cap H$. Since M is existentially closed, this automorphism can be realized as conjugation in M. Thus we can choose $b\in M$ such that $H^b=H$ but $H\cap G\neq(H\cap G)^b$. Clearly,

$$\operatorname{typ}(h_1, \dots, h_n)=\operatorname{typ}(h_1^b, \dots, h_n^b).$$

Choose new symbols $t_1, \dots, t_n$ and define the group $\langle H, t_1, \dots, t_n\rangle$ by requiring

$$\operatorname{typ}(t_1, \dots, t_n, h_1^b, \dots, h_n^b)=\operatorname{typ}(g_1, \dots, g_n, h_1, \dots, h_n),$$

so that $\langle t_1, \dots, t_n\rangle\cap H=G^b\cap H=(G\cap H)^b$. We can then form the amalgamated free product

$$L=M\mathop{*}_{H}\langle H, t_1, \dots, t_n\rangle,$$

so that $M\subseteq L$ and, in L,

$$M\cap\{t_1, \dots, t_n\}=H\cap\{t_1, \dots, t_n\}=(G\cap H)^b;$$

in particular, $G\cap\{t_1, \dots, t_n\}\subseteq(G\cap H)^b$. Finally, we can form the HNN-extension

$$\langle L, t \mid t^{-1}g_it=t_i,\quad t^{-1}h_it=h_i^b \ 1\leqslant i\leqslant n\rangle.$$

For each $t_i\in L$ such that $t_i\notin(G\cap H)^b$, the inequalities $t_i\neq g_1, \dots, t_i\neq g_n$ hold in L and insure that $t_i\notin G$. Thus there is a finite set of inequalities which are soluble in L and which insure that

$$\{t_1, \dots, t_n\}\cap G\subseteq(H\cap G)^b.$$

Thus we may choose $k_1, \dots, k_n$ and a in M such that $a^{-1}g_ia=k_i$, $a^{-1}h_ia=h_i^b$, and $K\cap G\subseteq(H\cap G)^b$, where $K=\{k_1, \dots, k_n\}$. Since $(G\cap H)^b=G^b\cap H$, we have that

$$G\cap K\subseteq(G\cap H)^b\cap(G\cap H)\subsetneqq G\cap H,$$

by choice of b. So we have $G^a=K$, $H^a=H$, and $|G\cap K|<|G\cap H|$, as required. □

The other lemma that we need in order to prove Theorem 2.1 is essentially a special case of the fact that any amalgam $A\cup B\cup C\cup D$ of groups such that $A\cap C=B\cap D=1$ can be embedded in a group.

Before stating the lemma, we introduce a little notation.

Definition For any groups G, H, K, L such that the groups $\langle G, H\rangle$ and $\langle K, L\rangle$ generated by G and H and by K and L, respectively, are defined, we write

$$\operatorname{typ}(G, H) = \operatorname{typ}(K, L)$$

if there exists an isomorphism $\theta: \langle G, H\rangle \to \langle K, L\rangle$ such that $G\theta = K$ and $H\theta = L$.

Note If G, H, K, L are finitely generated subgroups of an existentially closed group M, then $\operatorname{typ}(G, H) = \operatorname{typ}(K, L)$ if and only if there exists an element $a \in M$ such that $G^a = K$ and $H^a = L$. So, in the case as above, where $\equiv$ is an equivalence relation respected by conjugation in M, if $\operatorname{typ}(G, H) = \operatorname{typ}(K, L)$ and $G \equiv H$ then $K \equiv L$.

Lemma 2.3 *Let G, H, K, and L be isomorphic finitely generated subgroups of an existentially closed group M, such that $G \cap H = K \cap L = 1$. Then we can choose subgroups R, S, T, U of M such that*

$$\operatorname{typ}(G, H) = \operatorname{typ}(R, S) = \operatorname{typ}(S, T) = \operatorname{typ}(T, U)$$

$$\textit{and } \operatorname{typ}(K, L) = \operatorname{typ}(U, R).$$

Proof Suppose that $G = \langle g_1, \dots, g_n\rangle$, $H = \langle h_1, \dots, h_n\rangle$, $K = \langle k_1, \dots, k_n\rangle$ and $L = \langle l_1, \dots, l_n\rangle$. Choose distinct variables $r_1, \dots, r_n, s_1, \dots, s_n, t_1, \dots, t_n, u_1, \dots, u_n$, and define groups

$$A_1 = \langle \bar{r}_1, \dots, \bar{r}_n, \bar{s}_1, \dots, \bar{s}_n\rangle, \qquad A_3 = \langle \bar{t}_1, \dots, \bar{t}_n, \bar{u}_1, \dots, \bar{u}_n\rangle$$
$$A_2 = \langle \bar{s}_1, \dots, \bar{s}_n, \bar{t}_1, \dots, \bar{t}_n\rangle, \qquad A_4 = \langle \bar{u}_1, \dots, \bar{u}_n, \bar{r}_1, \dots, \bar{r}_n\rangle$$

by the rules

$$\begin{aligned}\operatorname{typ}(g_1, \dots, g_n, h_1, \dots, h_n) &= \operatorname{typ}(\bar{r}_1, \dots, \bar{r}_n, \bar{s}_1, \dots, \bar{s}_n)\\ &= \operatorname{typ}(\bar{s}_1, \dots, \bar{s}_n, \bar{t}_1, \dots, \bar{t}_n)\\ &= \operatorname{typ}(\bar{t}_1, \dots, \bar{t}_n, \bar{u}_1, \dots, \bar{u}_n)\\ \operatorname{typ}(k_1, \dots, k_n, l_1, \dots, l_n) &= \operatorname{typ}(\bar{u}_1, \dots, \bar{u}_n, \bar{t}_1, \dots, \bar{t}_n).\end{aligned}$$

Let $R = \langle \bar{r}_1, \dots, \bar{r}_n\rangle$, $S = \langle \bar{s}_1, \dots, \bar{s}_n\rangle$, $T = \langle \bar{t}_1, \dots, \bar{t}_n\rangle$, and $U = \langle \bar{u}_1, \dots, \bar{u}_s\rangle$. Since G, H, K, and L are all isomorphic, we can form the amalgamated free products

$$B_1 = A_1 \underset{S}{*} A_2 \quad \text{and} \quad B_2 = A_3 \underset{U}{*} A_4.$$

Then, by the normal-form theorem for free products, since

$$R \cap S = S \cap T = T \cap U = G \cap H = 1 = K \cap L = U \cap R,$$

R and T generate their free product in both B_1 and B_2. So we can form the amalgamated free product

$$C = B_1 \underset{\langle R,T \rangle}{*} B_2.$$

Since A_1, A_2, A_3, and A_4 are embedded in C, we can form the HNN-extension

$$\Big\langle M * C, \bar{x}, \bar{y}, \bar{z}, \bar{w} \;\Big|\; \langle G, H \rangle = \bar{x}^{-1}A_1\bar{x} = \bar{y}^{-1}A_2\bar{y} = \bar{z}^{-1}A_2\bar{z},$$
$$\langle K, L \rangle = \bar{w}^{-1}A_4\bar{w} \Big\rangle,$$

in which the equations

$$g_i = x^{-1}r_ix = y^{-1}s_iy = z^{-1}t_iz, \qquad k_i = w^{-1}u_iw$$
$$h_i = x^{-1}s_ix = y^{-1}t_iy = z^{-1}u_iz, \qquad l_i = w^{-1}r_iw$$

for $1 \leqslant i \leqslant n$, can all be solved. So we can choose the solutions in M, and these solutions generate groups R, S, T, and U as required. □

Proof of Theorem 2.1 Let Γ be the set of all subgroups of M which are isomorphic to G, and let $\equiv$ be some non-trivial equivalence relation on Γ which is respected by the action of M. As in the discussion above Lemma 2.2, we need to show that $H_1 \equiv H_2$, for all H_1 and H_2 in Γ.

Since $\equiv$ is not equality, we can choose G_1 and G_2 in Γ such that

$$|G_1 \cap G_2| < |G_1| \quad \text{and} \quad G_1 \equiv G_2.$$

If $|G_1 \cap G_2| \neq 1$, then, by Lemma 2.2, we can choose $G_3 \in \Gamma$ and $a \in M$ such that

$$G_1^a = G_3, \qquad G_2^a = G_2, \qquad |G_1 \cap G_3| < |G_1 \cap G_2|.$$

Since the action of M respects $\equiv$, we have that $G_1 \equiv G_3$. If $|G_1 \cap G_3| \neq 1$, then we use Lemma 2.2 again to find $G_4 \in \Gamma$ such that

$$|G_1 \cap G_4| < |G_1 \cap G_3| \quad \text{and} \quad G_1 \equiv G_4.$$

Since G_1 is finite, this process eventually stops, giving

$$|G_n \cap G_1| = 1 \quad \text{and} \quad G_n \equiv G_1.$$

Thus we can, and we will, suppose that $|G_1 \cap G_2| = 1$.

Given $H_1, H_2 \in \Gamma$ we can always choose $K \in \Gamma$ such that

$$|H_1 \cap K| = |H_2 \cap K| = 1.$$

To see this, consider the variables $x_1, \ldots, x_n$, where $n = |G|$. Then there are finitely many equations in $x_1, \ldots, x_n$ which will ensure that any solution $\bar{x}_1, \ldots, \bar{x}_n$ generates a homomorphic image of G. There are finitely many inequalities which will ensure that this homomorphic image is not a proper subgroup of G. For each i, there are $2n$ inequalities which ensure that $\bar{x}_i \notin H_1 \cup H_2$. The set $\mathscr{S}$ of all these equations and inequalities is finite, and solvable in $M * G$. Thus, we can choose solutions in M, and take K to be the group generated by these solutions. Clearly, if $H_1 \equiv K$ and $H_2 \equiv K$ then $H_1 \equiv H_2$, so we only need to show that $H_1 \equiv H_2$ for groups $H_1, H_2 \in \Gamma$ such that $|H_1 \cap H_2| = 1$.

By Lemma 2.3, if

$$|G_1 \cap G_2| = |H_1 \cap H_2| = 1,$$

we can find groups R_1, R_2, R_3, R_4 in Γ, such that

$$\begin{aligned}&\operatorname{typ}(G_1, G_2) = \operatorname{typ}(R_1, R_2) = \operatorname{typ}(R_2, R_3) = \operatorname{typ}(R_3, R_4)\\&\operatorname{typ}(H_1, H_2) = \operatorname{typ}(R_4, R_1).\end{aligned}$$

So, by the note above, $G_1 \equiv G_2$ implies $H_1 \equiv H_2$. □

Remark 1 If G is a finite subgroup of an existentially closed group M, and if G has a non-trivial characteristic subgroup H, then $\mathrm{N}_M(G)$ is not maximal in M. For, we have $\mathrm{N}_M(G) \subseteq \mathrm{N}_M(H)$ since H is characteristic in G, but $\mathrm{N}_M(G) \neq \mathrm{N}_M(H)$ because, for any $g_0 \in G \backslash H$, the set

$$\{x^{-1}hx = h, \quad x^{-1}g_0x \neq g \mid h \in H, g \in G\}$$

is soluble in the HNN-extension

$$\langle M, t \mid t^{-1}ht = h; \quad \forall h \in H \rangle.$$

Thus we have $\mathrm{N}_M(H) \not\subseteq \mathrm{N}_M(G)$. Of course, $\mathrm{N}_M(H) \neq M$, since the statement $x^{-1}Hx \neq H$ is equivalent to a finite set of solvable inequalities over M.

Remark 2 In the case where $G = \mathbb{Z}_2$, we can prove Theorem 2.1 using Theorem 1.4 in place of Lemma 2.3, as follows.

If a and b are elements of M of order 2, and if $\langle a \rangle \neq \langle b \rangle$, then $\langle a, b \rangle$ is dihedral: either D_∞ or D_{2n}, where $n > 1$ is the order of ab. Consider any $c \in M$ of order 2. We show that, if $\langle a \rangle \equiv \langle b \rangle$, then $\langle a \rangle \equiv \langle c \rangle$. We may suppose that $\langle a, c \rangle$ is not cyclic, since, in this case, $a = c$ and so $\langle a \rangle \equiv \langle c \rangle$ trivially.

Let ac have order $m > 1$. Then, as in the proof of Theorem 1.4, we can choose $r, s, t \in M$, such that rs and st have order n, and tr has order m. Then

$$\begin{aligned}&\operatorname{typ}(a, b) = \operatorname{typ}(r, s) = \operatorname{typ}(s, t),\\&\operatorname{typ}(a, c) = \operatorname{typ}(t, r).\end{aligned}$$

By the above note, $\langle a \rangle \equiv \langle b \rangle$ implies that $\langle r \rangle \equiv \langle s \rangle \equiv \langle t \rangle$, and hence that $\langle a \rangle \equiv \langle c \rangle$. So any non-trivial equivalence relation is universal, as required.

The point that we want to make here is that, for the case $G = \mathbb{Z}_2$, we can find three groups R_1, R_2, and R_3 in M such that

$$\operatorname{typ}(G_1, G_2) = \operatorname{typ}(R_1, R_2) = \operatorname{typ}(R_2, R_3),$$
$$\operatorname{typ}(H_1, H_2) = \operatorname{typ}(R_1, R_3),$$

but in the general case we need four groups R_1, R_2, R_3, and R_4. So we are prompted to put the following question, a positive answer to which would give a proof of this type for Theorem 2.1 for all G of the form $\mathbb{Z}_p$.

Question Is it true that, for any existentially closed group M and for any elements $a,b,c \in M$ of prime order p, there exist $x,y \in M$ such that $x^{-1}ax = b$, $y^{-1}ay = a$, and $x^{-1}bx = ycy^{-1}$?

Exercise Prove that the answer to the above Question depends only on p, and not on the particular existentially closed group.

The next lemma forms part of the proof of Theorem 2.5.

Lemma 2.4 *Let G be any group, and let $B_1, \ldots, B_k$ be any subgroups of G. Let L be the repeated* HNN*-extension*

$$\langle G, s_1, \ldots, s_k \mid s_i^{-1}bs_i = b; \quad \forall b \in B_i \ (1 \leqslant i \leqslant k)\rangle$$

of G. Then the subgroup $K \subseteq L$ generated by G and $t = (s_1s_2 \cdots s_k)^2$ is an HNN*-extension of G with t centralizing $\bigcap_{1 \leqslant i \leqslant k} B_i = B$. So*

$$K = \langle G, t \mid t^{-1}bt = b; \quad \forall b \in B\rangle.$$

Proof Let $X = \langle G, d \mid d^{-1}bd = b; \quad \forall b \in B\rangle$ be an HNN-extension. We want to show that the rules $\theta(x) = t$ and $\theta(g) = g$, $\forall g \in G$, define an isomorphism $\theta: X \to K$. Since t centralizes B, the rules do define a homomorphism, which is clearly surjective. It is easy to see that any element of X can be written in the form $w = g_0 d^{\varepsilon_1} g_1 \cdots d^{\varepsilon_n} g_n$, where $\varepsilon_i = \pm 1$, and if $\varepsilon_i = -\varepsilon_{i+1}$ then $g_i \notin B$. So we want to show that, if $\theta(w) = 1$, then $w = 1$, and by Britton's Lemma this means that $n = 0$.

So suppose that w is an element of X written in the above form, with $n \geqslant 1$. For $1 \leqslant i \leqslant n$, if $\varepsilon_i = -\varepsilon_{i+1} = 1$, then $g_i \notin B$, so we can choose $i_1 \leqslant k$ such that

$$g_i \in B_k \cap B_{k-1} \cap \cdots \cap B_{i_1+1}, \quad \text{but } g_i \notin B_{i_1}.$$

For each i, define r_i by:

$$r_i = \begin{cases} s_1 s_2 \cdots s_{i_1} & \text{if } \varepsilon_i = -\varepsilon_{i+1} = 1, \\ s_1 s_2 \cdots s_k & \text{otherwise.} \end{cases}$$

Let $r = s_1 s_2 \cdots s_k$, and consider $r^{\varepsilon_i} g_i r^{\varepsilon_{i+1}}$. If $\varepsilon_i = \varepsilon_{i+1}$, or if $\varepsilon_i = -1$, then

$r_i = r$. If $\varepsilon_i = -\varepsilon_{i+1} = 1$, then

$$r^{\varepsilon_i} g_i r^{\varepsilon_{i+1}} = r g_i r^{-1} = s_1 \cdots s_{i_1} g_i s_{i_1}^{-1} \cdots s_1^{-1} = r_i g_i r_i^{-1}.$$

So, in every case we have that

$$r^{\varepsilon_i} g_i r^{\varepsilon_{i+1}} = r_i^{\varepsilon_i} g_i r_i^{\varepsilon_{i+1}}.$$

Thus, we can rewrite $\theta(w)$ in the form

$$\theta(w) = g_0 r^{\varepsilon_1} r_1^{\varepsilon_1} g_1 r_1^{\varepsilon_2} r_2^{\varepsilon_2} g_2 r_2^{\varepsilon_3} \cdots r_{n-1}^{\varepsilon_{n-1}} g_{n-1} r_{n-1}^{\varepsilon_n} r^{\varepsilon_n} g_n.$$

Note that, if $r_i \neq r$, then $\varepsilon_i = 1$ and $\varepsilon_{i+1} = -1 \neq 1$. So, by definition, $r_{i+1} = r$. Similarly, since $\varepsilon_i \neq -1$, we have $r_{i-1} = r$.

Let $L_1 = \langle G, s_1, \dots, s_{k-1} \rangle \subseteq L$, and consider L as the HNN-extension

$$\langle L_1, s_k \mid s_k^{-1} b_k s_k = b_k; \quad \forall b_k \in B_k \rangle.$$

As a word in L, the element $\theta(w)$ has the form

$$\theta(w) = u_0 s_k^{\delta_1} u_1 s_k^{\delta_2} u_2 \cdots s_k^{\delta_m} u_m,$$

where $\delta_j = \pm 1$ and $u_j \in L_1$. (We have $m \leqslant 2n$; and $m \geqslant 1$, since $\theta(w) = g_0 r^{\varepsilon_1} v r^{\varepsilon_n} g_n$, for some v, and $n \geqslant 1$.) Hence by Britton's Lemma applied to $\langle L_1, s_k \rangle$, if $\theta(w) = 1$, then $\theta(w)$ must contain a subword of the form $s_k^{\delta} u_j s_k^{-\delta}$, with $u_j \in B_k$. Since $\langle s_1, \dots, s_k \rangle \cap G = \{1\}$, and since, as we have seen, if r_i does not involve s_k then $r_{i-1} = r_{i+1} = r$, we have that u_j must be of the form: either

$$s_k^{-1} s_{k-1}^{-1} \cdots s_1^{-1} g_i s_1 \cdots s_{k-1} s_k \tag{1}$$

or

$$s_k s_1 \cdots s_{i_1} g_i s_{i_1}^{-1} \cdots s_1^{-1} s_k^{-1}. \tag{2}$$

In the first case, since $u_j \in B_k$ and s_k centralizes B_k, we have

$$u_j = s_{k-1}^{-1} \cdots s_1^{-1} g_i s_1 \cdots s_{k-1} \in B_k \subseteq G.$$

Applying Britton's Lemma to $\langle G, s_1, \dots, s_{k-1} \rangle$, we must have $u_j \in B_{k-1}$, and hence

$$u_j = s_{k-2}^{-1} \cdots s_1^{-1} g_i s_1 \cdots s_{k-2} \in G.$$

Continuing the application of Britton's Lemma in this way to $\langle G, s_1, \dots, s_{k-q} \rangle$ $(1 \leqslant q \leqslant k)$, we see that

$$u_j \in B_k \cap B_{k-1} \cap \cdots \cap B_1$$

and $g_i = u_j$. So $g_i \in B$, which is a contradiction.

In the second case, the same type of argument shows that

$$g_i = u_j \in B_1 \cap B_2 \cap \cdots \cap B_{i_1}.$$

In this case, by definition of i_1,

$$g_i \in B_{i_1+1} \cap \cdots \cap B_k,$$

so again $g_i \in B$, contrary to assumption.

So we have that if $n \geqslant 1$ then $\theta(w) \neq 1$, and hence $\theta(w) = 1$ implies $n = 0$, $w = g_0$, and $\theta(w) = g_0 = w$. So $\theta(w) = 1$ implies $w = 1$ and $\langle G, t \rangle$ is an HNN-extension of the required type. $\square$

Note The above result is not always true if we take $t = s_1 s_2 \cdots s_k$. It may be possible to choose

$$g \in \bigcap_{2 \leqslant i \leqslant k} B_i, \qquad g \notin B_k, \qquad g \in \mathrm{N}_G(B_k), \qquad h \in B_1 \backslash B_2.$$

Then, $tgt^{-1}htg^{-1}t^{-1}ghg^{-1} = 1$. For example, let

$$G = \langle a, b \mid a^2 = b^2 = (ab)^2 = 1 \rangle,$$

let $B_1 = \langle a \rangle$, let $B_2 = \langle b \rangle$, let $g = b$, and let $h = a$. Then

$$\begin{aligned} tgt^{-1}htg^{-1}t^{-1}ghg^{-1} &= s_1 s_2 b s_2^{-1} s_1^{-1} a s_1 s_2 b s_2^{-1} s_1^{-1} a \\ &= s_1 b s_1^{-1} a s_1 b s_1^{-1} a \\ &= s_1 bab s_1^{-1} a \\ &= s_1 a s_1^{-1} a \\ &= 1. \end{aligned}$$

So $\langle G, t \rangle$ is not an HNN-extension in this way, and hence we need $t = (s_1 \cdots s_k)^2$.

Let $B_1, \ldots, B_k$ be subgroups of a group G. Then

$$\langle \mathrm{C}_G(B_i) \mid 1 \leqslant i \leqslant k \rangle \subseteq \mathrm{C}_G(B_1 \cap \cdots \cap B_k),$$

and this inclusion is usually proper. However, we have the following embedding theorem.

Theorem 2.5 *If $B_1, \ldots, B_k$ are subgroups of a group G, then there exists a group H, containing G, such that*

$$\langle \mathrm{C}_H(B_i) \mid 1 \leqslant i \leqslant k \rangle = \mathrm{C}_H(B_1 \cap \cdots \cap B_k).$$

Proof First we see that it is sufficient to show, for any $g \in \mathrm{C}_G(B_1 \cap \cdots \cap B_k)$, that there exists a group $L(g)$, containing G, such that $g \in \langle \mathrm{C}_{L(g)}(B_i) \mid 1 \leqslant i \leqslant k \rangle$, as follows. Form the free product G^* of all these $L(g)$, amalgamating G. So

$$\mathrm{C}_G(B_1 \cap \cdots \cap B_k) \subseteq \langle \mathrm{C}_{G^*}(B_i) \mid 1 \leqslant i \leqslant k \rangle.$$

Then define a sequence $G_0 \subseteq G_1 \subseteq G_2 \subseteq \ldots$, with $G_0 = G$ and $G_{j+1} = G_j^*$, and take

$$H = \bigcup_{j=0}^{\infty} G_j.$$

If $g \in C_H(B_1 \cap \cdots \cap B_k)$ then, for some j, $g \in C_{G_j}(B_1 \cap \cdots \cap B_k)$ and then

$$g \in \langle C_{G_{j+1}}(B_i) \mid 1 \leq i \leq k \rangle \subseteq \langle C_H(B_i) \mid 1 \leq i \leq k \rangle.$$

By Lemma 2.4, the group

$$L = \langle G, s_1, \ldots, s_k \mid s_i^{-1} b s_i = b; \quad \forall b \in B_i \ (1 \leq i \leq k) \rangle$$

contains the HNN-extension

$$K = \langle G, t \mid t^{-1} u t = u; \quad \forall u \in \bigcap_{1 \leq i \leq k} B_i = B \rangle,$$

with $t = (s_1 \cdots s_k)^2 \in \langle C_L(B_i) \mid 1 \leq i \leq k \rangle$. Now, for any $g \in C_G(B_1 \cap \cdots \cap B_k)$, the element gt centralizes B. So there is a homomorphism

$$\theta : K \rightarrow K,$$

which fixes the elements of G and maps t to gt, which is clearly surjective. We can write any non-trivial element of K in the form

$$g_0 t^{\varepsilon_1} g_1 t^{\varepsilon_2} g_2 \cdots t^{\varepsilon_n} g_n,$$

where $\varepsilon_i = \pm 1$, $g_i \in G$, and if $\varepsilon_j = -\varepsilon_{j+1}$ then $g_j \notin B$. Then if

$$g_0 (gt)^{\varepsilon_1} g_1 \cdots (gt)^{\varepsilon_n} g_n = 1,$$

by Britton's Lemma, for some j, we have $\varepsilon_j = -\varepsilon_{j+1}$ and g_j or $g^{-1} g_j g$ belongs to B. In either case, $g_j \in B$, which is a contradiction. So

$$g_0 (gt)^{\varepsilon_1} g_1 \cdots (gt)^{\varepsilon_n} g_n \neq 1$$

and θ is an automorphism. Thus we can form the HNN-extension

$$L(g) = \langle L, r \mid r^{-1} t r = gt, \quad r^{-1} x r = x; \quad \forall x \in G \rangle,$$

of L. Then r and t belong to $\langle C_{L(g)}(B_i) \mid 1 \leq i \leq k \rangle$, and hence so does g. Thus we have the required group $L(g)$. □

Lemma 2.6† *If $B_1, \ldots, B_k$ are finitely generated subgroups of the existentially closed group M, then*

$$\langle C_M(B_i) \mid 1 \leq i \leq k \rangle = C_M(B_1 \cap \cdots \cap B_k).$$

Proof Suppose that $g \in C_M(B_1 \cap \cdots \cap B_k)$. By Theorem 2.5, there is a group H, containing M, such that

$$g = x_{i_1} x_{i_2} \cdots x_{i_r}, \quad \text{where} \quad x_{i_j} \in C_H(B_j). \qquad (*)$$

Since each B_i is finitely generated, the statement (*) can be written as a

† For further results in this direction, see Hickin (1985).

finite set of equations over M, which are soluble in H and hence in M. So

$$g \in \langle \mathrm{C}_M(B_i) \mid 1 \leq i \leq k \rangle.$$

Obviously, $\langle \mathrm{C}_M(B_i) \mid 1 \leq i \leq k \rangle \subseteq \mathrm{C}_M(B_1 \cap \cdots \cap B_k)$, so we have the result. □

Theorem 2.7 *Let G be a finite subgroup of an existentially closed group M. Then:*

(a) If X is any subgroup of M containing $\mathrm{C}_M(G)$, there exists $H \leq G$ such that $\mathrm{C}_M(H) \subseteq X \subseteq \mathrm{N}_M(H)$.

(b) The set of subgroups of M containing $\mathrm{C}_M(G)$ is finite.

Proof Let $\mathcal{S} = \{A \mid A$ is a finitely generated subgroup of M and $\mathrm{C}_M(A) \subseteq X\}$. Now, $\mathcal{S}$ contains the finite group G, and thus the intersection of all the elements of $\mathcal{S}$ is a finite group H (say). Then H is equal to the intersection of finitely many of the members of $\mathcal{S}$ ($G \cap A$ is either G or strictly smaller than G, and G is finite, so eventually

$$G \cap A_2 \cap A_3 \cap \cdots \cap A_n$$

is smallest possible). Suppose that

$$H = A_1 \cap A_2 \cap \cdots \cap A_n.$$

Then, by Theorem 2.6, $H \in \mathcal{S}$, i.e. $\mathrm{C}_M(H) \subseteq X$. For any $x \in X$, we have

$$\mathrm{C}_M(x^{-1}Ax) = x^{-1}\mathrm{C}_M(A)x$$

and so $x^{-1}Ax \in \mathcal{S}$. Then $H \subseteq x^{-1}Hx$ and so $x \in \mathrm{N}_M(H)$. This proves part (a).

There are only finitely many subgroups of G. We show that, given a finite group H, there are only finitely many groups X such that $\mathrm{C}_M(H) \subseteq X \subseteq \mathrm{N}_M(H)$. We can then deduce (b) from (a).

If $X \subseteq \mathrm{N}_M(H)$ then X acts on H by conjugation and thus determines a unique subgroup of $\mathrm{Aut}\, H$. If $\mathrm{C}_M(H) \subseteq X \subseteq \mathrm{N}_M(H)$, and if

$$\mathrm{C}_M(H) \subseteq Y \subseteq \mathrm{N}_M(H),$$

where both X and Y determine the same subgroup of $\mathrm{Aut}\, H$, we have that, for each $y \in Y$, there exists an $x \in X$ such that $x^{-1}hx = y^{-1}hy$, for all $h \in H$. So $x^{-1}y \in \mathrm{C}_M(H) \subseteq X$ and thus $Y \subseteq X$. So $X = Y$, by a similar argument, and there is a one-to-one map from the set $\mathcal{X}$ of all groups satisfying $\mathrm{C}_M(H) \subseteq X \subseteq \mathrm{N}_m(H)$, into $\mathrm{Aut}\, H$. So, $\mathcal{X}$ is finite if H is finite. □

Remark We can deduce Theorem 2.1 from Theorem 2.7, as follows. Suppose that G is also characteristically simple, and suppose that $\mathrm{N}_M(G) \subseteq X \subseteq M$. Since $\mathrm{C}_M(G) \subseteq \mathrm{N}_M(G)$, by Theorem 2.7(a) there is a subgroup H of G such that $C_M(H) \subseteq X \subseteq N_M(H)$. Every automorphism of G is conjugation in some group containing G, and hence in M, and since $N_M(G) \subseteq N_M(H)$, H is fixed by every automorphism of G. Thus, if G is characteristically simple, then $H = 1$ or $H = G$. If $H = G$ then $X = N_M(H)$, and if $H = 1$ then $C_M(H) = M$, and so $X = M$. Thus, $N_M(G)$ is maximal in M.

Lemma 2.8 *Let B be a subgroup of a group G, let $k \in N_G(B) \backslash B$, and let $g \in C_G(B)$. Then there exists a group $H \supseteq G$ such that g is a product of conjugates of k and k^{-1} by elements of $C_H(B)$.*

Proof Let $\langle G, s \rangle$ be the HNN-extension of G with s centralizing B. Let

$$t = s^{-1}ks^2k^{-1}s^{-1}.$$

Since k normalizes B, it is easy to see that t centralizes B. So, to show that $\langle G, t \rangle$ is the HNN-extension

$$\langle G, t \mid t^{-1}bt = b; \quad \forall b \in B \rangle,$$

we see that, as in the proof of Lemma 2.4, it is sufficient to show that, if

$$w = g_0 t^{\varepsilon_1} g_1 t^{\varepsilon_2} g_2 \cdots t^{\varepsilon_n} g_n,$$

where $n \geq 1$ and where $g_i \in B$ only if $\varepsilon_i \neq -\varepsilon_{i+1}$, then $w \neq 1$ in $\langle G, t \rangle$.

Now,

$$w = g_0(s^{-1}ks^2k^{-1}s^{-1})^{\varepsilon_1} g_1 \cdots (s^{-1}ks^2k^{-1}s^{-1})^{\varepsilon_n} g_n;$$

so, applying Britton's Lemma to $\langle G, s \rangle$, we have that $w = 1$ only if w contains a subword of the form

$$s^{\delta} u s^{-\delta} \quad (\text{with } u \in B \text{ and } \delta = \pm 1). \tag{$*$}$$

Since $k \notin B$, we cannot have $u = k$. Thus, we have to have $\delta = -\varepsilon_i$ and $u = g_i$. But, by choice of w, if $g_i \in B$ then $-\varepsilon_{i+1} \neq \varepsilon_i = -\delta$; so w cannot contain a subword of the form ($*$) above, and hence $w \neq 1$.

As in the proof of Theorem 2.5, since $g \in C_G(B)$, there is an automorphism

$$\theta : \langle G, t \rangle \to \langle G, t \rangle,$$

which fixes G elementwise, and which maps t to gt. So we can form the HNN-extension

$$H = \langle G, s, r \mid r^{-1}tr = gt, \quad r^{-1}xr = x; \ \forall x \in G \rangle,$$

of $\langle G, s \rangle$. Then $r, t \in C_H(B)$ and $g = r^{-1}trt^{-1}$, giving the result. □

The proof of the next theorem requires a note and a corollary of Lemma 2.8.

Note Suppose that B is a finitely generated subgroup of an existentially closed group M. We can take a finite set of equations, in the variables x and y, which ensure that x and y centralize B. We can also take the inequality

$$x^{-1}y^{-1}xy \neq 1,$$

and the whole set can be solved in the direct product $M \times S_3$, where S_3 is the symmetric group on three letters (or indeed any non-Abelian group). Thus, we can choose elements in M that centralize B but do not commute. Thus $C_M(B)$ is not Abelian, and hence

$$C_M(B)/Z(B)$$

is non-trivial.

Corollary 2.9 *Suppose that B is a finitely generated subgroup of an existentially closed group M. Then, for any $k \in N_M(B)\backslash B$, any $g \in C_M(B)$ can be expressed as a product of conjugates of k and k^{-1} by elements of $C_M(B)$.*

Proof Let $B = \langle b_1, \ldots, b_n \rangle$. By Lemma 2.8, for any $g \in C_M(B)$ and for any $k \in N_M(B)\backslash B$, the set of equations

$$\{g = r^{-1}s^{-1}ks^2k^{-1}s^{-1}rsks^{-2}k^{-1}s, \quad s^{-1}b_is = b_i, \quad r^{-1}b_ir = b_i \mid 1 \leqslant i \leqslant n\}$$

can be solved over M. So the set can be solved in M and we can take $H = M$ in Lemma 2.8. □

Theorem 2.10 *If B is a finitely generated subgroup of an existentially closed group M, then the quotient group $C_M(B)/Z(B)$ is simple and is isomorphic to the unique minimal non-trivial normal subgroup of $N_M(B)/B$.*

Proof Suppose that there exists a group $X \neq Z(B)$ such that

$$Z(B) \subseteq X \trianglelefteq C_M(B).$$

Then $X\backslash Z(B) = X\backslash(B \cap C_M(B)) = X\backslash B$, and so $X\backslash B$ is non-empty. Thus we can find $x \in X \cap (N_M(B)\backslash B)$ and, by Corollary 2.9, every element of $C_M(B)$ is a product of conjugates of x and x^{-1} by elements of $C_M(B)$. Since $X \trianglelefteq C_M(B)$, we then have that $C_M(B) \subseteq X$ and so $X = C_M(B)$. Thus, $C_M(B)/Z(B)$ is simple.

It is easy to see that the homomorphism from $B.C_M(B)$ to $C_M(B)/Z(B)$ given by $cb \mapsto cZ(B)$, for all $b \in B$ and $c \in C_M(B)$, is well-defined and surjective, with kernel B. If $B \subsetneqq Y \trianglelefteq N_M(B)$, then we can take $y \in Y\backslash B$ and every element of $C_M(B)$ is a product of conjugates of y and y^{-1} by elements of $C_M(B)$. Since $C_M(B) \subseteq N_M(B)$ and $Y \trianglelefteq N_M(B)$, we have $C_M(B) \subseteq Y$ and so $(B.C_M(B))/B$ is the unique minimal non-trivial normal subgroup of $N_M(B)/B$. □

The next two theorems illustrate how Theorem 2.7 fails for subgroups of M which are not finite.

Theorem 2.11 *Let $G=\langle g\rangle$ be an infinite cyclic subgroup of an existentially closed group M. Let*

$$A=\bigcup_{i=1}^{\infty} C_M(g^i), \qquad B=\{x\in M \mid G\cap G^x\neq 1\}$$

($G^x=x^{-1}Gx$). Then A and B are subgroups of M, A is normal in B, B/A is isomorphic to the multiplicative group of the rationals $\mathbb{Q}^*$, B is a maximal subgroup of M, and A is simple.

Proof Clearly, both A and B contain the identity and are closed under inverses. If $x,y\in M$ each centralize some element of G, then, since G is cyclic, there is some element of G which they both centralize. Also, if $x^{-1}g^ix\in G$ and $y^{-1}g^jy\in G$, then $x^{-1}y^{-1}g^{ij}yx\in G$; so A and B are subgroups of M.

Suppose that $x\in B$. Then $x^{-1}g^rx=g^s$, for some $r,s\in\mathbb{Z}\backslash\{0\}$. If $x^{-1}g^nx=g^m$, then $g^{ns}=g^{mr}$ and thus $ns=mr$, so we can define a homomorphism $\theta:B\to\mathbb{Q}^*$ by the rule $x\theta=r/s$. For any $n,m\in\mathbb{Z}\backslash\{0\}$, the equation $y^{-1}g^ny=g^m$ can be solved in some HNN-extension of M (since g has infinite order) and thus in M. So θ is surjective. The kernel of θ is A, so A is normal in B, and B/A is isomorphic to $\mathbb{Q}^*$.

If $x\in M\backslash B$, then $G\cap G^x=\{1\}$, so

$$M=C_M(G\cap G^x)=\langle C_M(G), C_M(G^x)\rangle,$$

by Theorem 2.6. But

$$C_M(G^x)=x^{-1}C_M(G)x \quad\text{and}\quad C_M(G)\subseteq B,$$

so $M\subseteq\langle B, x\rangle$, and thus B is maximal in M.

For $x\in A\backslash\{1\}$, there exists some $i>0$ such that $x^{-1}g^ix=g^i$, and we can choose i large enough so that if $x=g^j$ then $i>|j|$. Let $G_k=\langle g^k\rangle$, then, if $k=in$ for some $n\geqslant 1$, we have $x\in C_M(G_k)$ and $x\notin G_k$. Since G_k is finitely generated, we can use Corollary 2.9 to deduce that $C_M(G_k)$ belongs to $\langle x\rangle^A$. For any $y\in A$, there exists $p\neq 0$ such that $y^{-1}g^py=g^p$. If we take $k=ip$, we get that $y\in C_M(G_k)$, and so $y\in\langle x\rangle^A$. Thus, A is the normal subgroup of A generated by x, and so A is simple. □

Theorem 2.12 *Let G be a free Abelian group, of rank n, in an existentially closed group M. Let*

$$\Gamma=\{H\leqslant G \mid \text{rank } H=n\},$$

let

$$A=\bigcup\nolimits_{H\in\Gamma} C_M(H),$$

and let

$$B=\{x\in M \mid G\cap G^x\in\Gamma\}.$$

Then A and B are subgroups of M, A is normal in B, A is simple, the quotient group, B/A, is isomorphic to $\mathrm{GL}(n, \mathbb{Q})$, *and B is maximal in M.*

Proof Let D be a free Abelian group, then, if D has rank n, any subgroup of D is free Abelian of rank at most n. If $D_1 \leq D$ is of rank n, then, for all $d \in D$, there exists some $m \geq 1$ such that $d^m \in D_1$. For otherwise $\langle d \rangle \cap D_1 = \{1\}$, and $\langle D_1, d \rangle = D_1 \times \langle d \rangle$, which has rank $n+1$. Thus, if $D_2 \leq D$ also has rank n, the intersection $D_1 \cap D_2$ contains non-trivial powers of every generator of D, and hence $\mathrm{rank}\,(D_1 \cap D_2) = n$.

Thus Γ is closed under intersection, and so A is closed under multiplication. Clearly A is closed under inverses.

Since $G \cap G^{xy^{-1}} = y(G^y \cap G^x)y^{-1}$, we have

$$\mathrm{rank}\,(G \cap G^{xy^{-1}}) = \mathrm{rank}\,(G^y \cap G^x).$$

Since $G^x \cap G^y \cap G = (G^x \cap G) \cap (G^y \cap G)$, we see from the above remarks that, if $x, y \in B$, then $\mathrm{rank}\,(G^x \cap G^y \cap G) = n$. But

$$n = \mathrm{rank}\, G^x \geq \mathrm{rank}\,(G^x \cap G^y) \geq \mathrm{rank}\,(G^x \cap G^y \cap G),$$

so $\mathrm{rank}\,(G^x \cap G^y) = n$, and hence $xy^{-1} \in B$.

Let $G = \langle g_1, \ldots, g_n \rangle$. If $x \in B$, then $\mathrm{rank}\,(G \cap G^{x^{-1}}) = n$, so there exists an integer $r \geq 1$ such that $g_i^r \in G \cap G^{x^{-1}}$ $(1 \leq i \leq n)$. Define

$$\theta : B \to \mathrm{GL}\,(n, \mathbb{Q})$$

by:

$$x\theta = (s_{ij}/r)_{i,j},$$

where

$$g_i^r = x g_1^{s_{i1}} g_2^{s_{i2}} \cdots g_n^{s_{in}} x^{-1}.$$

If $(m_1, m_2, \ldots, m_n)(s_{ij})_{i,j} = (0, 0, \ldots, 0)$, then

$$g_1^{m_1 r} g_2^{m_2 r} \cdots g_n^{m_n r} = 1,$$

and so, since the g_i are free Abelian generators for G, we have $m_1 = m_2 = \cdots = m_n = 0$. Thus, $\det\,(s_{ij})_{i,j} \neq 0$ and so $x\theta \in \mathrm{GL}\,(n, \mathbb{Q})$.

If

$$g_i^q = x g_1^{t_{i1}} g_2^{t_{i2}} \cdots g_n^{t_{in}} x^{-1},$$

then

$$x g_1^{t_{i1} r} \cdots g_n^{t_{in} r} x^{-1} = g_i^{qr} = x g_1^{s_{i1} q} \cdots g_n^{s_{in} q} x^{-1},$$

and so $t_{ij} r = s_{ij} q$ $(1 \leq i, j \leq n)$. Thus θ is well defined.

If

$$g_i^q = y g_1^{t_{i1}} g_2^{t_{i2}} \cdots g_n^{t_{in}} y^{-1} \quad (1 \leq i \leq n)$$

then $g_i^{rq} \in G \cap G^{xy}$ and

$$g_i^{rq} = xy g_1^{\alpha_{i1}} g_2^{\alpha_{i2}} \cdots g_n^{\alpha_{in}} y^{-1} x^{-1},$$

where $\alpha_{ij} = \sum_{1 \leq k \leq n} s_{ik} t_{kj}$. Thus $x\theta \cdot y\theta = (xy)\theta$.

If $x \in \ker \theta$, then $x^{-1}g_i^r x = g_i^r$ $(1 \leqslant i \leqslant n)$. So x centralizes $\langle g_1^r, \ldots, g_n^r \rangle$, which has rank n, and $x \in A$. If $x \in A$, then $x \in C_M(H)$, where $H \in \Gamma$, and $H \subseteq G \cap G^x$. Since H has rank n, there exists an integer $m \geqslant 1$ such that $g_i^m \in H$ $(1 \leqslant i \leqslant m)$. Then

$$xg_i^m x^{-1} = g_i^m \quad (1 \leqslant i \leqslant m),$$

and so $x\theta = I_n$. Thus $A = \ker \theta$.

Any matrix P in $\mathrm{GL}(n, \mathbb{Q})$ can be written in the form $(s_{ij}/r)_{i,j}$, for some $r \geqslant 1$ and $s_{ij} \in \mathbb{Z}$. Since $\det P \neq 0$, we have

$$\operatorname{rank} \langle g_1^r, \ldots, g_n^r \rangle = n = \operatorname{rank} \langle g_1^{s_{11}} \cdots g_n^{s_{1n}}, \ldots, g_1^{s_{n1}} \cdots g_n^{s_{nn}} \rangle,$$

and thus there is an isomorphism

$$\langle g_1^r, \ldots, g_n^r \rangle \to \langle g_1^{s_{11}} \cdots g_n^{s_{1n}}, \ldots, g_1^{s_{n1}} \cdots g_n^{s_{nn}} \rangle,$$

which carries g_i^r to $g_1^{s_{i1}} g_2^{s_{i2}} \cdots g_n^{s_{in}}$ $(1 \leqslant i \leqslant n)$. Thus, the equations

$$g_i^r = z g_1^{s_{i1}} g_2^{s_{i2}} \cdots g_n^{s_{in}} z^{-1} \quad (1 \leqslant i \leqslant n),$$

can be solved in an HNN-extension of M, and hence in M. Thus, $\bar{z}\theta = P$, θ is surjective, and so

$$B/A \cong \mathrm{GL}(n, \mathbb{Q}).$$

Suppose that $x \in M \backslash B$, so that $G \cap G^x$ has rank at most $n-1$. Let $H_0 = G \cap G^x$, and choose $h_1, \ldots, h_m$ in G such that

$$H_0 = \langle h_1 \rangle \times \langle h_2 \rangle \times \cdots \times \langle h_m \rangle.$$

Since rank $H_0 < n$, we can choose i_1 such that

$$H_0 \cap \langle g_{i_1} \rangle = \{1\}.$$

By re-ordering, we will suppose that $i_1 = m+1$. If $H_1 = \langle H_0, g_{m+1} \rangle$ then

$$H_1 = H_0 \times \langle g_{m+1} \rangle,$$

and if rank $H_1 < n$, we can choose i_2 such that

$$H_1 \cap \langle g_{i_2} \rangle = \{1\}.$$

Again we may suppose $i_2 = m+2$, and

$$H_2 = \langle H_1, g_{m+2} \rangle = H_1 \times \langle g_{m+2} \rangle.$$

Carrying on in this way, we eventually get the group

$$H = H_{m-n} = H_0 \times \langle g_{m+1} \rangle \times \cdots \times \langle g_n \rangle$$

which is an element of Γ. For $1 \leqslant j \leqslant m$, interchanging h_j and g_n, and fixing the other generators of H, defines an automorphism $\theta_j : H \to H$. Since, for all triples D_1, D_2, D_3 of pairwise disjoint Abelian groups,

$(D_1 \times D_2) \cap (D_1 \times D_3) = D_1$, we have

$$H_0 \cap H_0\theta_1 \cap \cdots \cap H_0\theta_m = \{1\}.$$

Since M is existentially closed and H is finitely generated, we can choose $y_j \in \mathrm{N}_M(H)$ such that $y_j^{-1}h_l y_j = h_l\theta_j$ $(1 \leqslant j \leqslant m,\ 1 \leqslant l \leqslant m)$. Then,

$$\begin{aligned} M = \mathrm{C}_M(H_0 \cap H_0^{y_1} \cap \cdots \cap H_0^{y_m}) &= \langle \mathrm{C}_M(H_0), \mathrm{C}_M(H_0^{y_1}), \ldots, \mathrm{C}_M(H_0^{y_m}) \rangle \\ &\subseteq \langle \mathrm{C}_M(H_0), y_1, \ldots, y_m \rangle \\ &= \langle \mathrm{C}_M(G), \mathrm{C}_M(G^x), y_1, \ldots, y_m \rangle \\ &\subseteq \langle \mathrm{C}_M(G), x, y_1, \ldots, y_m \rangle. \end{aligned}$$

Also, $H = H^{y_j} \subseteq G \cap G^{y_j}$, and H has rank n. So, $G \cap G^{y_j}$ has rank n, and $y_j \in B$. Of course, $\mathrm{C}_M(G) \subseteq A \subseteq B$, so

$$M \subseteq \langle B, x \rangle \subseteq M,$$

and thus B is a maximal in M.

If $x \in A$, then there exists some $r \geqslant 1$ such that $x^{-1}g_i^r x = g_i^r$ $(1 \leqslant i \leqslant n)$. If $x = g_1^{s_1} \cdots g_n^{s_n}$, then choose $r > |s_i|$. Then, for any k which is divisible by r, we have $x \in \mathrm{C}_M(G_k)$, where $G_k = \langle g_1^k, \ldots, g_n^k \rangle$. Since $x \notin G_k$, then $\mathrm{C}_M(G_k) \subseteq \langle x \rangle^{\mathrm{C}_M(G_k)}$. But G_k has rank n, so $\mathrm{C}_M(G_k) \subseteq A$, and thus $\mathrm{C}_M(G_k) \subseteq \langle x \rangle^A$. For any $y \in A$, we can find s such that $y^{-1}g_i^{rs}y = g_i^{rs}$ $(1 \leqslant i \leqslant n)$. Hence $y \in \mathrm{C}_M(G_{sr}) \subseteq \langle x \rangle^A$. So $A = \langle x \rangle^A$, and A is simple.

□

Remark There is an analogue of Theorem 2.12 for free groups, in which the role previously played by the subgroups of rank n is played by the subgroups of G of finite index in G. However, it is not clear how to identify the quotient group, B/A. We do not know how general such a theorem can be made.

3
ω-Homogeneous groups

We now set up the necessary machinery to prove, at the end of the chapter, that there does indeed exist an existentially closed group. We will consider classes $\mathscr{X}$ of groups, and conditions SC, JEP, AEP, and AC, which these classes may satisfy. We then prove a sequence of results, all of the same basic type, the last of which is that, if $\mathscr{X}$ is non-trivial and contains only countably many isomorphism types of groups, then $\mathscr{X}$ is the skeleton of an existentially closed group if and only if it satisfies SC, JEP, and AC. This allows us to prove the main theorem of this chapter, which is that there exists a locally finitely presented countable existentially closed group. This group is one of an infinite family of existentially closed groups which is discussed in more detail in Chapter 4.

Definition 3.1 A group K is said to be *ω-homogeneous* if, given any finite set $\{g_1, \ldots, g_r, h\} \subseteq K$ and any monomorphism

$$\theta : \langle g_1, \ldots, g_r \rangle \to K,$$

we can extend θ to a monomorphism $\varphi : \langle g_1, \ldots, g_r, h \rangle \to K$. (We see below that in a countable group, this is equivalent to saying that given any two finite subsets of the same type there is an automorphism of K which maps one to the other.)

Theorem 3.2 *An existentially closed group M is ω-homogeneous.*

Proof There exists an HNN-extension of M in which θ is equivalent to conjugation. Thus θ is equivalent to conjugation in M, and hence can be extended to an inner automorphism of M. □

Examples 1. If A is Abelian and contains an element of infinite order, then A is ω-homogeneous if and only if A is divisible. If A is an Abelian torsion group, write $A = A_1 \oplus A_2 \oplus \cdots$, where A_i is a p_i-group, for some prime p_i, and $p_i \neq p_j$ if $i \neq j$. Then A is ω-homogeneous if and only if each A_i is ω-homogeneous. And A_i is ω-homogeneous if and only if it is either divisible or a direct sum of cyclic groups of the same order.

2. If A is Abelian and ω-homogeneous, and contains no involutions,

then the group $\langle A, t \mid t^2 = 1,\ t^{-1}at = a^{-1};\ \forall a \in A\rangle$ is also ω-homogeneous.

3. If G and H are finite ω-homogeneous groups, with co-prime orders, then their direct product $H \times G$ is ω-homogeneous.

4. There are a few other finite ω-homogeneous groups, for example A_4 and A_5.

5. P. Hall's universal locally finite group $\bigcup_{i\in\mathbb{N}} S_{n_i}$, where $n_0 \geqslant 3$, $n_{i+1} = (n_i)!$, and S_{n_i} is embedded in $S_{n_{i+1}}$ by means of the regular representation, is ω-homogeneous.

Definition 3.3 For any group G, the *skeleton* of G, denoted $\operatorname{Sk} G$, is the class of all finitely generated groups that can be embedded in G.

Theorem 3.4 (i) *If G is a countable group and if K is an ω-homogeneous group with $\operatorname{Sk} G \subseteq \operatorname{Sk} K$, then G can be embedded in K.*

(*ii*) *Two countable ω-homogeneous groups are isomorphic if and only if they have the same skeleton.*

(*iii*) *Any isomorphism between finitely generated subgroups of a countable ω-homogeneous group K can be extended to an automorphism of K.*

Proof (i) Let $G = \{g_1, g_2, \ldots\}$, and let

$$G_n = \langle g_1, g_2, \ldots, g_n\rangle.$$

We show that we can construct embeddings

$$\varphi_n : G_n \to K,$$

such that φ_{n+1} extends φ_n $(n \geqslant 1)$. Then the map

$$\varphi : G \to K$$

which is defined by the rule

$$g_n\varphi = g_n\varphi_n \quad (n \geqslant 1),$$

is an embedding of G in K.

Since G_n is finitely generated, and since $\operatorname{Sk} G \subseteq \operatorname{Sk} K$, there exists an embedding

$$\theta_n : G_n \to K.$$

Let $\varphi_1 = \theta_1$.

We define φ_{n+1} inductively. Suppose that $\varphi_n : G_n \to K$ extends $\varphi_{n-1} : G_{n-1} \to K$. The map

$$(\theta_{n+1}|_{G_n})^{-1} \circ \varphi_n$$

is an embedding of $\langle g_1\theta_{n+1}, \ldots, g_n\theta_{n+1}\rangle$ into K, which carries $g_i\theta_{n+1}$ to

$g_i\varphi_n$ $(1 \leqslant i \leqslant n)$. Since K is ω-homogeneous, this map can be extended to an embedding ψ_{n+1} of Im θ_{n+1} into K. Putting

$$\varphi_{n+1} = \theta_{n+1} \circ \psi_{n+1},$$

we see that

$$g_i\varphi_{n+1} = g_i\theta_{n+1}\theta_{n+1}^{-1}\varphi_n = g_i\varphi_n \quad (1 \leqslant i \leqslant n).$$

So φ_{n+1} does extend φ_n, as required.

(ii) Suppose that $K_1 = \langle g_1, g_2, \ldots \rangle$ and $K_2 = \langle h_1, h_2, \ldots \rangle$ are ω-homogeneous groups which have the same skeleton. Let $G_1 = \langle g_1 \rangle$ and choose an embedding $\theta_1 : G_1 \rightarrow K_2$, as we can, since Sk $K_1 =$ Sk K_2. Then the group $\langle g_1\theta_1, h_1 \rangle = H_1$ belongs to Sk K_2 and hence to Sk K_1, so there is an embedding $\varphi_1 : H_1 \rightarrow K_1$. Since K_1 is ω-homogeneous, we can choose φ_1 so that $g_1\theta_1\varphi_1 = g_1$. Let $G_2 = \langle H_1\varphi_1, g_2 \rangle$. Then there is an embedding $\theta_2 : G_2 \rightarrow K_2$ and, since K_2 is ω-homogeneous, we can choose θ_2 to extend θ_1 and so that $h_1\varphi_1\theta_2 = h_1$. Let $H_2 = \langle G_2\theta_2, h_2 \rangle$. Continuing in this way, we choose φ_i to extend φ_{i-1} and so that $g_i\theta_i\varphi_i = g_i$, and then choose θ_{i+1} to extend θ_i and such that $h_i\varphi_i\theta_{i+1} = h_i$. Thus we can define embeddings $\theta : K_1 \rightarrow K_2$ and $\varphi : K_2 \rightarrow K_1$, by $g_i\theta = g_i\theta_i$ and $h_i\varphi = h_i\varphi_i$ $(i \in \mathbb{N})$, so that $\theta\varphi = \varphi\theta = 1$. Thus K_1 and K_2 are isomorphic.

(iii) This follows by an obvious modification of the argument in (ii), putting $K_1 = K_2 = K$, taking θ_1 to be the given isomorphism and G to be its domain, and then extending θ_1 to an automorphism of K, as in (ii). □

Theorem 3.5 *If M is a countable existentially closed group, and if G is a finitely generated subgroup of M with $\mathrm{Z}(G) = 1$, then $\mathrm{C}_M(G)$ is isomorphic to M.*

Proof We show that $\mathrm{C}_M(G)$ is ω-homogeneous and that Sk $\mathrm{C}_M(G) =$ Sk M. The result then follows by Theorem 3.4(ii).

Clearly, Sk $\mathrm{C}_M(G) \subseteq$ Sk M. If F is a finitely generated subgroup of M, take a group $F_1 \cong F$ and form the direct product $M \times F_1$. Then form the HNN-extension $\langle M \times F_1, t \mid t^{-1}Ft = F_1 \rangle$, so that $t^{-1}Ft$ centralizes G. Since M is existentially closed, and F and G are finitely generated, we can find $x \in M$ such that $x^{-1}Fx \subseteq \mathrm{C}_M(G)$. So Sk $M =$ Sk $\mathrm{C}_M(G)$.

If H is a finitely generated subgroup of $\mathrm{C}_M(G)$ and if $\theta : H \rightarrow \mathrm{C}_M(G)$ is a monomorphism, then, since $\mathrm{Z}(G) = 1$, we have $H \cap G = H\theta \cap G = 1$. So $\langle H, G \rangle \cong H \times G$, and $\langle H\theta, G \rangle \cong H\theta \times G$. Thus, θ extends to a monomorphism $\hat{\theta} : H \times G \rightarrow M$, which fixes G elementwise. Since G and H are finitely generated, there exists $k \in M$ such that, for all $x \in H \times G$, $k^{-1}xk = x\hat{\theta}$. But $k \in \mathrm{C}_M(G)$, so θ is the restriction of an inner automorphism of $\mathrm{C}_M(G)$. Hence, $\mathrm{C}_M(G)$ is ω-homogeneous. □

Remark Deeper results of a similar nature have been proved (see Hickin 1985); for example, it has been shown that any existentially closed group is equivalent to an existentially closed group of the same power which is embeddable in a maximal subgroup of itself.

We shall now look at classes of groups, and properties which are sufficient to make a class the skeleton of an existentially closed group.

Notation A class $\mathscr{X}$ of groups will always be isomorphism-closed, and will be called *trivial* if it consists precisely of the groups with one element.

Definition 3.6 A class $\mathscr{X}$ of finitely generated groups is said to satisfy:

(i) *SC* (subgroup closure) if, whenever $F \in \mathscr{X}$ and G is a finitely generated subgroup of F, then $G \in \mathscr{X}$.

(ii) *JEP* (the joint embedding property) if, for any $F, G \in \mathscr{X}$, there exists a group $H \in \mathscr{X}$ and monomorphisms θ and φ such that $\theta: F \rightarrow H$ and $\varphi: G \rightarrow H$.

(iii) *AEP* (the amalgamated embedding property) if, for any $F, G, H \in \mathscr{X}$, and for any monomorphisms $\alpha: F \rightarrow G$ and $\beta: F \rightarrow H$, there exist $K \in \mathscr{X}$ and monomorphisms $\gamma: G \rightarrow K$ and $\delta: H \rightarrow K$ such that $\alpha\gamma = \beta\delta$.

(iv) *AC* (algebraic closure) if, whenever $F \in \mathscr{X}$ and $\mathscr{S}$ is a finite set of equations defined over F and soluble over F, then $\mathscr{S}$ is soluble in some group $G \in \mathscr{X}$ that contains F.

Theorem 3.7 *Let $\mathscr{X}$ be a non-empty class of finitely generated groups, which contains at most a countable set of isomorphism types of groups. Then $\mathscr{X}$ is the skeleton of a countable group if and only if it satisfies* SC *and* JEP.

Proof If $\mathscr{X} = \operatorname{Sk} K$, where K is a countable group, then $\mathscr{X}$ clearly satisfies SC. If $F, G \in \mathscr{X}$, then $F \cong F_1 \subseteq K$, and $G \cong G_1 \subseteq K$. Let H be the subgroup of K generated by F_1 and G_1. Then H is finitely generated, and there exist monomorphisms

$$\theta: F \rightarrow F_1 \subseteq H, \qquad \varphi: G \rightarrow G_1 \subseteq H.$$

So $\mathscr{X}$ satisfies JEP.

Suppose that $\mathscr{X}$ satisfies SC and JEP. Let $G_0, G_1, \ldots$ be representatives of the isomorphism classes in $\mathscr{X}$. Define $H_0 = G_0$, and define H_{i+1} to be a group in $\mathscr{X}$ in which both H_i and G_{i+1} are embedded. Then, identifying H_i with its image in H_{i+1}, we form $K = \bigcup_{i \in \mathbb{N}} H_i$. Since each H_i is countable, K is countable. Every G_i is embedded in K, so $\mathscr{X} \subseteq \operatorname{Sk} K$. If F is a finitely generated subgroup of K, then, for some $i \in \mathbb{N}$, we have $F \subseteq H_i \in \mathscr{X}$. So $\operatorname{Sk} K = \mathscr{X}$ and K is the required group. □

Theorem 3.8 *Let $\mathscr{X}$ be a non-empty class of finitely generated groups* *containing only a countable set of isomorphism types. Then $\mathscr{X}$ is the skeleton of a countable ω-homogeneous group if and only if it satisfies* SC *and* AEP.

Proof However we write out the proof of this result, it seems complicated. So we begin by outlining the 'plan of attack'.

First we prove the 'only if' part of the theorem, and then we prove that, if any non-empty class satisfies SC and AEP, it satisfies JEP. So that $\mathscr{X} = \operatorname{Sk} K_0$, for some countable group K_0. The proof then proceeds in three stages.

I. We show that, for any finitely generated group $G \subseteq K_0$, any $h \in K_0$, and any monomorphism $\theta : G \to K_0$, there is a countable group $\hat{K}_0$, containing K_0, and a monomorphism $\hat{\theta} : \langle G, h \rangle \to \hat{K}_0$, such that $\hat{\theta}$ extends θ and $\operatorname{Sk} \hat{K}_0 = \mathscr{X}$.

II. We then show that there exists a countable group X_0, containing K_0, such that, for all finitely generated groups $G \subseteq K_0$, elements $h \in K_0$, and monomorphisms $\theta : G \to K_0$, there is a monomorphism $\hat{\theta} : \langle G, h \rangle \to X_0$ that extends θ. We show, further, that we can choose X_0 such that $\mathscr{X} = \operatorname{Sk} X_0$.

III. Finally, we show that there exists a countable group X, with $\mathscr{X} = \operatorname{Sk} X$, such that for each finitely generated group $G \subseteq X$, element $h \in X$, and monomorphism $\theta : G \to X$, we can find a monomorphism $\hat{\theta} : \langle G, h \rangle \to X$, which extends θ. So this X is the required ω-homogeneous group.

If $\mathscr{X} = \operatorname{Sk} X$, where X is a countable ω-homogeneous group, then $\mathscr{X}$ clearly satisfies SC. If $F, G, H \in \mathscr{X}$, there exist isomorphisms

$$\theta : F \to F_1 \subseteq X, \quad \varphi : G \to G_1 \subseteq X, \quad \psi : H \to H_1 \subseteq X.$$

If $\alpha : F \to G$ and $\beta : F \to H$ are embeddings, then the inverse of the embedding $\theta^{-1}\beta\psi : F_1 \to H_1 \subseteq X$ can be extended to an embedding $\delta_1 : H_1 \to X$, since X is ω-homogeneous. Similarly, the inverse of the embedding $\theta^{-1}\alpha\varphi : F_1 \to G_1 \subseteq X$ can be extended to an embedding $\gamma_1 : G_1 \to X$. Putting $K = \langle H_1\delta_1, G_1\gamma_1 \rangle \subseteq X$, $\delta = \psi\delta_1$ and $\gamma = \varphi\gamma_1$, we get $\delta : H \to K$, $\gamma : G \to K$, and $\alpha\gamma = \theta\,(\theta^{-1}\alpha\varphi\gamma_1) = \theta = \theta(\theta^{-1}\beta\psi\delta_1) = \beta\delta$. So $\mathscr{X}$ satisfies AEP.

Now suppose that $\mathscr{X}$ satisfies SC and AEP. Since $\mathscr{X}$ is non-empty and satisfies SC, we have $1 \in \mathscr{X}$. Given any $H, G \in \mathscr{X}$, putting $F = 1$, there exist monomorphisms $\alpha : F \to H$ and $\beta : F \to G$. Since $\mathscr{X}$ satisfies AEP there exists $K \in \mathscr{X}$ such that G and H can be embedded in K. Thus $\mathscr{X}$ satisfies JEP. So, by Theorem 3.7, there exists a countable group, K_0, with $\mathscr{X} = \operatorname{Sk} K_0$.

Let $G \subseteq K_0$ be finitely generated, let $h \in K_0$, and let $\theta : G \to K_0$ be a monomorphism. We now do stage I. Since K_0 is countable, we can write

K_0 as $\bigcup_{i\in\mathbb{N}} G_i$, where each G_i is finitely generated, and

$$G, G\theta, \{h\} \subseteq G_0 \subseteq G_1 \subseteq G_2 \subseteq \cdots.$$

Thus, we get a sequence of embeddings

$$\begin{array}{ccccccccccccc} G & \xrightarrow{\theta} & G_0 & \xrightarrow{1} & G_1 & \xrightarrow{1} & G_2 & \xrightarrow{1} & \cdots & \xrightarrow{1} & G_i & \xrightarrow{1} & G_{i+1} \xrightarrow{1} \cdots \\ {\scriptstyle 1}\big\downarrow & & & & & & & & & & & & \\ G_0. & & & & & & & & & & & & \end{array}$$

Since $\mathscr{X}$ satisfies AEP, we can find a finitely generated group $\hat{G}_0$ and embeddings $\alpha_0: G_0 \to \hat{G}_0$ and $\beta_0: G_0 \to \hat{G}_0$ such that $\theta\alpha_0 = \beta_0$. By induction, for each $i \in \mathbb{N}$, we can find a finitely generated group $\hat{G}_i$, and embeddings $\alpha_i: G_i \to \hat{G}_i$ and $\beta_i: \hat{G}_{i-1} \to \hat{G}_i$, such that $\alpha_{i-1}\beta_i = \alpha_i$. Thus we can extend the above diagram, using AEP, to a commuting ladder

$$\begin{array}{ccccccccccc} G & \xrightarrow{\theta} & G_0 & \xrightarrow{1} & G_1 & \xrightarrow{1} \cdots \xrightarrow{1} & G_i & \xrightarrow{1} & G_{i+1} & \xrightarrow{1} & \cdots \\ {\scriptstyle 1}\big\downarrow & & {\scriptstyle \alpha_0}\big\downarrow & & {\scriptstyle \alpha_1}\big\downarrow & & {\scriptstyle \alpha_i}\big\downarrow & & {\scriptstyle \alpha_{i+1}}\big\downarrow & & \\ G_0 & \xrightarrow[\beta_0]{} & \hat{G}_0 & \xrightarrow[\beta_1]{} & \hat{G}_1 & \xrightarrow[\beta_2]{} \cdots \xrightarrow[\beta_i]{} & \hat{G}_i & \xrightarrow[\beta_{i+1}]{} & \hat{G}_{i+1} & \xrightarrow[\beta_{i+2}]{} & \cdots, \end{array}$$

where $\hat{G}_i \in \mathscr{X}$ and $\alpha_i\beta_{i+1} = \alpha_{i+1}$. But, since $\mathscr{X}$ is isomorphism-closed, we may choose $\hat{G}_i$ so that $\alpha_i = 1$. Then $\beta_i = 1$, for $i \geqslant 1$, and θ is the restriction of β_0 to G. Take $\hat{\theta}$ to be the restriction of β_0 to $\langle G, h\rangle \subseteq G_0$, and put $\hat{K}_0 = \bigcup_{i\in\mathbb{N}} \hat{G}_i$. Then $\hat{\theta}: \langle G, h\rangle \to \hat{K}_0$ is a monomorphism which extends θ, $\hat{K}_0$ is countable, $\mathscr{X} \subseteq \operatorname{Sk} \hat{K}_0$ since $K_0 \subseteq \hat{K}_0$, and each $\hat{G}_i \in \mathscr{X}$; so $\operatorname{Sk} \hat{K}_0 \subseteq \mathscr{X}$. Thus, we have completed stage I.

There are only countably many triples (G, h, θ) with $G \subseteq K_0$ finitely generated, $h \in K_0$, and $\theta: G \to K_0$ a monomorphism. We may list these triples (G_0, h_0, θ_0), (G_1, h_1, θ_1), (G_2, h_2, θ_2), ... Now, construct $\hat{K}_0 = K_1$ as in stage I, so that $\theta_0: G_0 \to K_0$ extends to $\hat{\theta}_0: \langle G_0, h_0\rangle \to K_1$. Then $\theta_1: G_1 \to K_0 \subseteq K_1$; thus we can construct $\hat{K}_1 = K_2$, so that θ_1 extends to $\hat{\theta}_1: \langle G_1, h_1\rangle \to K_2$. In this fashion, we construct a sequence of groups $K_0 \subseteq K_1 \subseteq K_2 \subseteq \cdots$, each of which is countable and such that $\operatorname{Sk} K_i = \mathscr{X}$, for all i. Let $X_0 = \bigcup_{i\in\mathbb{N}} K_i$; then X_0 is countable. Further, for each finitely generated group $G \subseteq K_0$, and each $h \in K_0$, each monomorphism $\theta: G \to K_0$ extends to a monomorphism $\hat{\theta}: \langle G, h\rangle \to K_i \subseteq X_0$ and, since every finitely generated subgroup of X_0 lies in some K_i, we have $\mathscr{X} = \operatorname{Sk} X_0$. Thus we have completed stage II.

Now, X_0 is countable and so there are only countably many triples (G, h, θ) such that $G \subseteq X_0$ is finitely generated, $h \in X_0$, and $\theta: G \to X_0$ is a monomorphism. So, as in stage II, we can construct a countable group X_1, with $\operatorname{Sk} X_1 = \mathscr{X}$ and such that, for any triple (G, h, θ), as above, the monomorphism $\theta: G \to X_0$ extends to a monomorphism $\hat{\theta}: \langle G, h \rangle \to X_1$. We can do the same with X_1, to get a countable group X_2, with $\operatorname{Sk} X_2 = \mathscr{X}$ and such that, for each $G \subseteq X_1$ and each $h \in X_1$, each monomorphism $\theta: G \to X_1$ extends to a monomorphism $\hat{\theta}: \langle G, h \rangle \to X_2$. In this way, we can construct a sequence $X_0 \subseteq X_1 \subseteq X_2 \subseteq \cdots$ of countable groups, such that $\mathscr{X} = \operatorname{Sk} X_i$ for each i. Let $X = \bigcup_{i \in \mathbb{N}} X_i$. Then X is countable, $\operatorname{Sk} X = \mathscr{X}$, and if $G \subseteq X$ is a finitely generated group, if $k \in X$, and if $\theta: G \to X$ is any monomorphism, then there exists i such that G, h, and $G\theta$ all lie in X_i. By construction, there exists a monomorphism $\hat{\theta}: \langle G, h \rangle \to X_{i+1}$, which extends $\theta: G \to X_i$. Thus, X is ω-homogeneous and we have completed stage III. □

Theorem 3.9 *If $\mathscr{X}$ is a non-trivial non-empty class of finitely generated groups that consists of at most a countable set of distinct isomorphism types, then $\mathscr{X}$ is the skeleton of a countable existentially closed group if and only if it satisfies* SC, JEP, *and* AC.

Proof The skeleton of any group must satisfy SC and JEP. If $\mathscr{X} = \operatorname{Sk} M$, where M is a countable existentially closed group, and if $F \in \mathscr{X}$, then F is isomorphic to some subgroup F_1 of M. If $\mathscr{S}$ is a finite set of equations over F, let $\mathscr{S}_1$ be the corresponding set of equations over F_1. If $\mathscr{S}$ is soluble over F then $\mathscr{S}_1$ is soluble over F_1, in G_1 (say). Then $\mathscr{S}_1$ is soluble over M (in $M *_{M \cap G_1} G_1$). Thus $\mathscr{S}_1$ is soluble in M and hence in a finitely generated subgroup H_1 of M containing F_1. Let H be a group, containing F, that is isomorphic to H_1. Then $\mathscr{S}$ is soluble in H, and $H \in \mathscr{X}$. So $\mathscr{X}$ satisfies AC.

Now we show that, if $\mathscr{X}$ satisfies SC, JEP, and AC then $\mathscr{X} = \operatorname{Sk} M$, for some countable existentially closed group M.

First, we show that if $\mathscr{X}$ satisfies JEP and AC then $\mathscr{X}$ satisfies AEP.

Suppose that $F, G, H \in \mathscr{X}$, and that $\alpha: F \to G$ and $\beta: F \to H$ are monomorphisms. Since $\mathscr{X}$ satisfies JEP, there exist $K_1 \in \mathscr{X}$ and monomorphisms $\theta: G \to K_1$ and $\varphi: H \to K_1$. Now, $F\alpha\theta$ and $F\beta\varphi$ are isomorphic subgroups of K_1 and so the finite set of equations

$$\{x^{-1}(f_i\alpha\theta)x = f_i\beta\varphi \mid 1 \leq i \leq n\},$$

where $F = \langle f_1, \ldots, f_n \rangle$, is soluble in some HNN-extension of K_1. Since $\mathscr{X}$ satisfies AC, there is thus some group $K \in \mathscr{X}$ which contains K_1 and an element k such that $k^{-1}(f_i\alpha\theta)k = f_i\beta\varphi$ $(1 \leq i \leq n)$. Hence the map $\rho: G\theta \to K$, given by $G\theta\rho = k^{-1}G\theta k$, is a monomorphism which carries $F\alpha\theta$ to $F\beta\varphi$. Let $\gamma = \theta\rho$ and $\delta = \varphi$. Then γ and δ are monomorphisms and, for each $f \in F$, we have

$$f\alpha\gamma = f\alpha\theta\rho = k^{-1}(f\alpha\theta)k = f\beta\varphi = f\beta\delta.$$

So $\mathscr{X}$ does satisfy AEP.

By Theorem 3.8, we have that there exists a countable ω-homogeneous group K such that $\mathscr{X} = \mathrm{Sk}\, K$.

If $\mathscr{S}$ is a finite set of equations over K, then $\mathscr{S}$ is also a set of equations over some finitely generated subgroup F of K. Since $\mathscr{X}$ satisfies AC, if $\mathscr{S}$ is soluble over K, and hence over F, there is some group $G \in \mathscr{X}$ that contains F, such that $\mathscr{S}$ is soluble in G. Now, G is isomorphic to some $G\theta \subseteq K$ and, since K is ω-homogeneous, the monomorphism $(\theta|_F)^{-1}: F\theta \to K$ extends to a monomorphism $\hat{\theta}: G\theta \to K$, so that $\theta\hat{\theta}$ fixes F elementwise. Since $G \cong G\theta\hat{\theta}$, the set $\mathscr{S}$ is soluble in $G\theta\hat{\theta}$ and hence in K. Thus K is algebraically closed. But $K \neq 1$, since $\mathscr{X}$ contains some non-trivial group; thus, by Theorem 1.7, K is existentially closed. □

We are now in a position to show that existentially closed groups do indeed exist.

Theorem 3.10 *There exists a locally finitely presented countable existentially closed group.*

Proof Let $\mathscr{X}$ be the class of finitely generated subgroups of finitely presented groups. Clearly, $\mathscr{X}$ satisfies SC. If $A, B \in \mathscr{X}$, let G and H be finitely presented groups containing A and B, respectively. The direct product $X = A \times B$ is a subgroup of $G \times H$, which is finitely presented. Thus $X \in \mathscr{X}$, and there are natural embeddings $\theta: A \to X$ and $\psi: B \to X$; so X satisfies JEP. If A is a finitely generated subgroup of the finitely presented group G, and if $\mathscr{S}$ is a finite set of equations over A, that are soluble in $L \supseteq A$, let

$$H = \langle A, h_1, \dots, h_n \mid w(\boldsymbol{h}) = 1 \quad \text{for all } w(\boldsymbol{x}) \in \mathscr{S} \rangle.$$

Since $\mathscr{S}$ is soluble in L, there is a homomorphism $\theta: H \to L$ given by $a\theta = a$, $\forall a \in A$, and $h_i\theta = \bar{x}_i$ $(1 \leqslant i \leqslant n)$, where $\{\bar{x}_1, \dots, \bar{x}_n\} \subseteq L$ is a set of solutions for $\mathscr{S}$. Since $A \leqslant L$, the map $\theta|_A$ is an embedding, and so A is a subgroup of H. We form the free product $F = H *_A G$ of H and G amalgamating A. Then F is finitely generated and, as defining relations for F, we can take a finite set Γ of defining relations for G, together with the defining relations for A as a subgroup of H, the set $\mathscr{S}$, and a finite set Δ of relations saying that the generating elements of A in G are identified with the corresponding elements of A in H. But the defining relations of A are consequences of $\Gamma \cup \Delta$ and so, as defining relations for F, we can take the finite set $\Gamma \cup \mathscr{S} \cup \Delta$. So $F \supseteq H$ is finitely presented; hence $H \in \mathscr{X}$ and thus $\mathscr{X}$ satisfies AC. There are, up to isomorphism, only countably many finitely presented groups, and each finitely presented group has only countably many finitely generated subgroups. So $\mathscr{X}$ contains at most countably many isomorphism types. Thus, by Theorem 3.9, there exists a countable existentially closed group M such that $\mathrm{Sk}\, M = \mathscr{X}$.

We must now show that M is locally finitely presented

If A is a finitely generated subgroup of M, then $A \in \mathscr{X}$; so there exists a finitely presented group G with $A \subseteq G$. But $G \in \mathscr{X}$, so G is isomorphic to some subgroup G_1 of M. If $A_1 \subseteq G_1$ is isomorphic to A, then $y^{-1}A_1y = A$ in some HNN-extension of M. Thus, y can be chosen in M, and A is a subset of $y^{-1}G_1y$, which is a finitely presented subgroup of M. Since M is countable, it is equal to the union of its finitely generated subgroups, and hence, by the above, to the union of its finitely presented subgroups. Thus, the finitely presented subgroups of M form a local system, and M is a countable locally finitely presented existentially closed group. □

4
Recursion theory

In this chapter we collect together the ideas and results from recursion theory that we shall need, and discuss ways in which they are used in group theory.

No prior knowledge of recursion theory is required, but if the reader wants formal definitions and proofs of the ideas and theorems discussed, he should refer to a standard text on recursive function theory, for example Rogers (1967).

Later in the chapter, we shall show that there are $2^{\aleph_0}$ countable existentially closed groups and we shall define the notion of Ziegler reducibility, which is fundamental to the rest of the discussion. However, we begin with some basic recursion theory.

Intuitively, a *recursive function* from the set of natural numbers $\mathbb{N}$ to itself is a function that can be calculated mechanically. A subset X of $\mathbb{N}$ is *recursive* if its characteristic function ($f(n)=1$ if $n \in X$, $f(n)=0$ if $n \notin X$) is recursive. A subset $X \subseteq \mathbb{N}$ is said to be *recursively enumerable,* usually abbreviated to r.e., if there exists a machine which constructs a list of numbers, one by one, and X is the set of those numbers on the list. The construction may go on for ever, and in a sense it must if X is infinite. Thus, for a recursively enumerable set X, if some number n belongs to X, we will know this sooner or later, since it will eventually appear on the list, but, if n does not belong to X, we will never discover this from the machine's output alone. (Of course, we may discover it by examining the way the machine is constructed, e.g. it may be that all those numbers less than or equal to n that lie in X must have been produced by the mth stage of the machine's calculation). To make these intuitive ideas precise, we would need to explain exactly what a machine is. We shall not do this here. For positive statements we shall rely on the principle (commonly called Church's Thesis) that the definition of a machine is sufficiently wide to allow any obviously mechanical process to be carried out by some machine within the definition. For negative results, it will be sufficient to make the assumption that the set of all machines can be mechanically enumerated; this will be explained more precisely later.

As we have described it so far, recursion theory exists on the underlying set $\mathbb{N}$. If, however, W is any effectively enumerated set, that is, any set for which there exists a mechanically calculable injective function $f: W \to \mathbb{N}$, then we can translate recursion theoretic ideas into W. For instance, $X \subseteq W$ is a recursively enumerable subset of W if $f(X)$ is a recursively enumerable subset of $\mathbb{N}$. It follows from Church's Thesis that the set of recursively enumerable subsets of W is independent of the choice of f. To see this, suppose that $f,g: W \to \mathbb{N}$ are mechanically calculable functions, and suppose that $f(X)$ is recursively enumerable. Let M be a machine which calculates $f(X)$, so that M produces $n_1, n_2, n_3, \ldots$ and

$$\{n_1, n_2, n_3, \ldots\} = f(X).$$

Form a new machine M' from M as follows: Set M going. Then, for each n_i produced by M, find $w_i \in W$ such that $f(w_i) = n_i$, (we can do this since f is mechanically calculable, so we just calculate $f(w)$ for $w \in W$ until we find w_i). Then calculate $g(w_i)$, and let M' produce the output $g(w_i)$.

So M' produces output

$$\{g(w_1), g(w_2), \ldots\},$$

and, since

$$\{f(w_1), f(w_2), \ldots\} = f(X),$$

we must have

$$\{w_1, w_2, \ldots\} = X,$$

since f is injective, and thus

$$\{g(w_1), g(w_2), \ldots\} = g(X).$$

By Church's Thesis, M' is a machine; so $g(X)$ is recursively enumerable.

The sets $\mathbb{N} \times \mathbb{N}$ and

$$\mathcal{P}_{\mathrm{f}}(\mathbb{N}) = \{X \subseteq \mathbb{N} \mid |X| < \infty\}$$

are both effectively enumerable, for example, by

$$f(n, m) = 2^n(2m+1) - 1 \quad \text{and} \quad g(X) = \sum_{n \in X} 2^n,$$

respectively. So we can talk about recursively enumerable subsets of $\mathcal{P}_{\mathrm{f}}(\mathbb{N})$ or $\mathbb{N} \times \mathbb{N}$.

As we have said, we will not give formal definitions of 'recursive' or 'recursively enumerable' sets, but other recursion-theoretic ideas that we shall need will be precisely defined in terms of these notions. Important among these are certain of the recursion-theoretic reducibilities. These are ways of making formal the intuitive ideas 'if X were recursive then Y would be' or 'if X were recursively enumerable then Y would be'. For recursive reducibilities, Y is reducible to X if, whenever the membership problem for X is solvable, so is the membership problem for Y. First we define the strongest of these reducibilities.

Definition 4.1 For $X, Y \subseteq \mathbb{N}$, *$Y$ is one–one reducible to X,* written $Y \leqslant_1 X$, if there exists an injective recursive function $f:\mathbb{N}\to\mathbb{N}$ such that $n \in Y$ if and only if $f(n) \in X$. If we wish to emphasize the function f, we shall say that $Y \leqslant_1 X$ *via* f.

It is clear that the relation $\leqslant_1$ is transitive, i.e. if $Y \leqslant_1 X$ and $X \leqslant_1 Z$ then $Y \leqslant_1 Z$, and so $\leqslant_1$ generates an equivalence relation, denoted by $\equiv_1$ and called 1–1 *equivalence.* The significance of this relation is shown in the following theorem.

Theorem 4.2 *For $X, Y \subseteq \mathbb{N}$, the relation $X \equiv_1 Y$ holds if and only if there is a recursive permutation of $\mathbb{N}$ which maps X to Y.*

If the inverse of a recursive function exists, then it is recursive. So, if π is a recursive permutation of $\mathbb{N}$ and $X\pi = Y$, then $X \leqslant_1 Y$ via π and $Y \leqslant_1 X$ via π^{-1}; hence $X \equiv_1 Y$. Thus, to prove Theorem 4.2, we need to show that, if $X \equiv_1 Y$, then such a π exists. The following lemma is sufficient for this.

Lemma 4.3 *If f and g are recursive injective maps from $\mathbb{N}$ to $\mathbb{N}$ then, whenever $X \leqslant_1 Y$ via f and $Y \leqslant_1 X$ via g, there exists a recursive permutation π of $\mathbb{N}$ such that $X\pi = Y$.*

Proof If $\theta:\mathbb{N}\to\mathbb{N}$ is any injective map, then, for any $n \in \mathbb{N}$, either $\theta^i(n) = \theta^j(n)$ for some $i<j$, in which case $\theta^i(n) = \theta^i(\theta^{j-i}(n))$ and so $n = \theta^{j-i}(n)$, or $\theta^i(n) \neq \theta^j(n)$ for $0 \leqslant i < j$. If $\theta^k(n) \neq n$ for any $k \geqslant 0$, then either $\theta^{-k}(n)$ is defined for all $k \geqslant 0$, or, for some $k \geqslant 0$, $\theta^{-k-1}(n)$ is not defined. Thus we see that any orbit of $\mathbb{N}$ under θ is one of the three following types:

I. *Finite,* of the form $(m, \theta(m), \dots, \theta^k(m))$,

II. *Infinite,* of the form

$$(\dots, \theta^{-i}(m), \dots, \theta^{-1}(m), m, \theta(m), \dots, \theta^i(m), \dots),$$

III. *Semifinite,* of the form $(m, \theta(m), \dots, \theta^i(m), \dots)$.

If $\mathbf{O}_{gf}(n)$ is an orbit of $\mathbb{N}$ under gf, then $\mathbf{O}_{fg}(f(n))$ is an orbit under fg, which is of the same type as $\mathbf{O}_{gf}(n)$ and, if this type is type I, then both orbits have the same length. (In the case of type III orbits, if

$$\mathbf{O}_{gf}(n) = (m, gf(m), (gf)^2(m), \dots)$$

we may have

$$\mathbf{O}_{fg}(f(n)) = (f(m), fgf(m), \dots)$$

or

$$\mathbf{O}_{fg}(f(n)) = (g^{-1}(m), f(m), fgf(m), \dots),$$

depending on whether or not $g^{-1}(m)$ is defined. But, in both cases, $\mathbf{O}_{fg}(f(n))$ is of type III.) In this way, we have a natural type-preserving one–one correspondence between the set of orbits of $\mathbb{N}$ under gf and the set of orbits of $\mathbb{N}$ under fg.

We construct a recursive permutation π of $\mathbb{N}$ which carries each gf-orbit to the corresponding fg-orbit.

Construct an increasing sequence $A_{-1} \subseteq A_0 \subseteq A_1 \subseteq \cdots$ of finite sets of pairs $(m, n) \in \mathbb{N} \times \mathbb{N}$ such that:

(i) No A_i contains both (m, n_1) and (m, n_2), if $n_1 \neq n_2$, or both (m_1, n) and (m_2, n), if $m_1 \neq m_2$. (This ensures that π is well defined and injective.)

(ii) If $(m, n) \in A_i$, then $\mathbf{O}_{gf}(m)$ and $\mathbf{O}_{fg}(n)$ correspond. (This ensures that if $X \leqslant_1 Y$ via f and $Y \leqslant_1 X$ via g, then $X\pi = Y$; see later.)

(iii) For $i \geqslant 1$, the set A_i contains the pairs (i, n) and (m, i), for some m and n. (This ensures that π maps $\mathbb{N}$ onto itself.)

Take A_{-1} to be the empty set $\varnothing$ and construct A_{i+1} from A_i as follows. First search to see if $(i, n) \in A_i$ for any n. If not, then add $(i, (fg)^k f(i))$, for the smallest integer $k \geqslant 0$ such that $(m, (fg)^k f(i)) \notin A_i$ for any m, if such a k exists. Then search to see if $(m, i) \in A_i$, for any m. If not, and if $i \neq (fg)^k f(i)$, then add $((gf)^h g(i), i)$ to A_i, for the smallest integer $h \geqslant 0$ such that $((gf)^h g(i), n) \notin A_i$ for any n, if such an h exists. Then A_{i+1} is taken to be A_i, with $(i, (fg)^k f(i))$ and $((gf)^h g(i), i)$ added if necessary. Now we observe that, if $(i, n) \notin A_i$ for any n, then there must exist an integer $k \geqslant 0$ such that $(m, (fg)^k f(i)) \notin A_i$ for any m, even if $\mathbf{O}_{fg}(f(i))$ is finite, as follows. If $\mathbf{O}_{fg}(f(i))$ is finite, $\mathbf{O}_{gf}(i)$ is finite of the same length. All the elements of A_i are of the form $(j, (fg)^q f(j))$, with $q \in \mathbb{Z}$, and so, if $(m, n) \in A_i$, then $m \in \mathbf{O}_{gf}(i)$ if and only if $n \in \mathbf{O}_{fg}(f(j))$. Since each integer can only appear once as the right member, and once as the left member, of a pair in A_i, it follows that, if $(i, n) \notin A_i$, then, for some k, we have $(m, (fg)^k f(i)) \notin A_i$. Similarly, if $(m, i) \notin A_i$ for any m, then there exists an h such that $((gf)^h g(i), n) \notin A_i$, for any n. So each A_i satisfies (i), (ii), and (iii).

We define π to be $\bigcup_{i=-1}^{+\infty} A_i$. Clearly, π satisfies (i), (ii), and (iii) and, by Church's Thesis, is recursive. If $X, Y \subseteq \mathbb{N}$ are any sets such that $X \leqslant_1 Y$ via f and $Y \leqslant_1 X$ via g, then X is a union of complete orbits of $\mathbb{N}$ under gf, and Y is the union of the corresponding orbits of $\mathbb{N}$ under fg. Thus we have constructed π so that $X\pi = Y$, and π is a recursive permutation of $\mathbb{N}$. □

Remark The equivalence classes under $\equiv_1$ form a partially ordered set, under the order induced by $\leqslant_1$. (So $[X] \leqslant [Y]$ if and only if $X \leqslant_1 Y$.) This partially ordered set is, in fact, an upper semilattice, the least upper bound to $[X]$ and $[Y]$ (the classes containing X and Y, respectively) being the class $[X \vee Y]$ containing the disjoint union $X \vee Y$, where $\vee$ is defined as follows.

Definition 4.4 For $X, Y \subseteq \mathbb{N}$ the *disjoint union* $X \vee Y$ is the set $\{2x \mid x \in X\} \cup \{2y+1 \mid y \in Y\} \subseteq \mathbb{N}$.

Here, of course, we have split $\mathbb{N}$ up into two infinite complementary recursive subsets (the odd and the even numbers), we have enumerated these sets as $\{a_i \mid i \in \mathbb{N}\}$ and $\{b_i \mid i \in \mathbb{N}\}$, we have mapped X into one of these sets by $i \mapsto a_i$, mapped Y into the other set by $i \mapsto b_i$, and we have taken the union of the images. It is clear from Theorem 4.2 that the equivalence class obtained is independent of the particular pair of complementary recursive subsets chosen: we have to make a particular choice but which choice we make is irrelevant. A similar remark applies to the other equivalence relations that we shall define in connection with other recursion-theoretic reducibilities.

Obviously, the map $f: \mathbb{N} \to \mathbb{N}$ given by $f(n) = 2n$ is recursive, and $n \in X$ if and only if $f(n) \in X \vee Y$. So $X \leqslant_1 X \vee Y$, and, similarly, $Y \leqslant_1 X \vee Y$. If $X \leqslant_1 Z$ via f and $Y \leqslant_1 Z$ via g, define $h: \mathbb{N} \to \mathbb{N}$ by $h(2n) = f(n)$ and $h(2n+1) = g(n)$. Then h is recursive, since f and g are recursive, and $m \in X \vee Y$ if and only if $m = 2n+1$ and $n \in Y$, or $m = 2n$ and $n \in X$, which is if and only if $m = 2n+1$ and $g(\frac{1}{2}(m-1)) \in Z$, or $m = 2n$ and $f(\frac{1}{2}m) \in Z$. So $m \in X \vee Y$ if and only if $h(m) \in Z$, and thus $X \vee Y \leqslant_1 Z$. Hence $[X \vee Y]$ is a least upper bound for $[X]$ and $[Y]$, as claimed.

Now, let $x_1, \ldots, x_r, x_1^{-1}, \ldots, x_r^{-1}$ be distinct symbols, and let W be the set of all words on these letters. If $a_1, \ldots, a_r$ are elements of a group, G, we map W into G by the three rules:

(i) $x_i \mapsto a_i \quad (1 \leqslant i \leqslant r)$,
(ii) $x_i^{-1} \mapsto a_i^{-1} \quad (1 \leqslant i \leqslant r)$,
(iii) If $u \mapsto g$ and $v \mapsto h$ then $uv \mapsto gh$.

Then by

$$\mathrm{Rel}\,(a_1, \ldots, a_r)$$

we mean the set of all those words in W which map to the identity of G. (We shall usually think of $\mathrm{Rel}\,(a_1, \ldots, a_r)$ as a set of words in the a_i and their inverses. But, for the purposes of formal definition, it is better not to do so. For example, if $a_1 = a_2$ we still require x_1 and x_2 to be distinct symbols, so that x_1 and x_2 are distinct members of $\mathrm{Rel}\,(a_1, \ldots, a_r)$.)

Theorem 4.5 *If* $b_1, \ldots, b_s$ *are elements of a group generated by* $a_1, \ldots, a_r$, *then* $\mathrm{Rel}\,(b_1, \ldots, b_s) \leqslant_1 \mathrm{Rel}\,(a_1, \ldots, a_r)$.

Proof We let $y_1, \ldots, y_s, y_1^{-1}, \ldots, y_s^{-1}, x_1, \ldots, x_r, x_1^{-1}, \ldots, x_r^{-1}$ be distinct symbols, and we take $\mathrm{W}(\boldsymbol{x})$ to be the set of words on the x's, and $\mathrm{W}(\boldsymbol{y})$ to be the set of words on the y's. Then $\mathrm{Rel}\,(b_1, \ldots, b_s) \subseteq \mathrm{W}(\boldsymbol{y})$, $\mathrm{Rel}\,(a_1, \ldots, a_r) \subseteq \mathrm{W}(\boldsymbol{x})$, and each of $\mathrm{W}(\boldsymbol{x})$ and $\mathrm{W}(\boldsymbol{y})$ is an effectively enumerable set. We require an injective recursive function

$$f: \mathrm{W}(\boldsymbol{y}) \to \mathrm{W}(\boldsymbol{x})$$

such that $z \in \text{Rel}(b_1, \dots, b_s)$ if and only if $f(z) \in \text{Rel}(a_1, \dots, a_r)$.

For any word $w \in W(\boldsymbol{x})$, we denote the formal inverse of w by w^*, that is, the word obtained from w by replacing each x_i by x_i^{-1} and each x_i^{-1} by x_i, and then reversing the result. Since $b_j \in \langle a_1, \dots, a_r\rangle$, there is a word $w_j \in W(\boldsymbol{x})$ such that $b_j = w_j(a_1, \dots, a_r)$. We would like to define a map from $W(\boldsymbol{y})$ to $W(\boldsymbol{x})$ in which $f(z)$ is obtained from $z(y_1, \dots, y_s)$ by replacing each y_i in z by w_i, and each y_i^{-1} by w_i^*. But, although this map takes elements of $\text{Rel}(b_1, \dots, b_s)$ to elements of $\text{Rel}(a_1, \dots, a_r)$, it may fail to be injective (e.g. if $b_1^{-1} = b_2$, we could have chosen w_2 to be w_1^*, in which case $y_1y_1^{-1} \mapsto w_1w_1^*$, and $y_1y_2 \mapsto w_1w_1^*$). So we make a slightly more complicated definition of f. We can enumerate the elements of $W(\boldsymbol{y})$, as $u_0, u_1, u_2, \dots$, in such a way that if $i \neq j$ then $u_i \neq u_j$. We can also effectively enumerate, as $v_0, v_1, v_2, \dots$, the subset of $W(\boldsymbol{x})$ that consists of words on $x_1, \dots, x_r$ only, and not their inverses, again so that, if $i \neq j$, then $v_i \neq v_j$. Now, if $z(y_1, \dots, y_s) \in W(\boldsymbol{y})$, we write

$$z(w_1, \dots, w_s) \quad (\in W(\boldsymbol{x}))$$

for the result of replacing each occurrence of y_i or y_i^{-1} in z by w_i or w_i^*, respectively. Define $f: W(\boldsymbol{y}) \to W(\boldsymbol{x})$ by the rule: $f(u_j) = u_j(w_1, \dots w_s)v_j^*v_j$ for $j \geqslant 1$ (we can, and we will, assume that u_0 and v_0 are the empty words of $W(\boldsymbol{y})$ and $W(\boldsymbol{x})$, respectively, and that $f(u_0) = v_0$). Then f is recursive and, since

$$\begin{aligned} u_j(b_1, \dots, b_s) &= u_j(w_1(\boldsymbol{a}), \dots, w_s(\boldsymbol{a})) \\ &= u_j(w_1(\boldsymbol{a}), \dots, w_s(\boldsymbol{a}))v_j^*(\boldsymbol{a})v_j(\boldsymbol{a}), \end{aligned}$$

(where $w_i(\boldsymbol{a}) = w_i(a_1, \dots, a_r)$), it is clear that $u_j \in \text{Rel}(b_1, \dots, b_s)$ if and only if $f(u_j) \in \text{Rel}(a_1, \dots, a_r)$. Now, if $f(u_j) = f(u_i)$, then $u_j(w_1, \dots, w_s)v_j^*v_j = u_i(w_1, \dots, w_s)v_i^*v_i$. We can suppose that the length of v_i is no less than the length of v_j (where the length of each x_i is 1 and the length of ux_i is the length of u plus 1). If $j = 0$, then u_j and v_j are the empty words, so $u_i(w_1, \dots, w_s)v_i^*v_i$ is empty and so $i = 0$. If $j \geqslant 1$ then, since v_i contains no symbols of the form x_k^{-1} and v_j^* contains only symbols of this form, to have $f(u_i) = f(u_j)$ we must have $v_i = v_j$, and hence $i = j$. Thus f is injective and

$$\text{Rel}(b_1, \dots, bs) \leqslant_1 \text{Rel}(a_1, \dots, a_r) \text{ via } f. \qquad \square$$

Corollary 4.6 *If* $\langle a_1, \dots, a_r\rangle = \langle b_1, \dots, b_s\rangle$ *then*

$$\text{Rel}(a_1, \dots, a_r) \equiv_1 \text{Rel}(b_1, \dots, b_s).$$

If G is a finitely generated group, we shall denote by

$$\text{Rel}\, G$$

any subset of $\mathbb{N}$ which is 1–1 equivalent to some $\text{Rel}(a_1, \dots, a_r)$, for some set $\{a_1, \dots, a_r\}$ of generators of G. By Corollary 4.6, $\operatorname{Rel} G$ is uniquely defined up to a recursive permutation of $\mathbb{N}$. Recursion-theoretic concepts are invariant under the group of recursive permutations, so that, in recursion-theoretic contexts, we may treat $\operatorname{Rel} G$ as a well defined subset of $\mathbb{N}$. It is only in recursion-theoretic contexts that we shall use $\operatorname{Rel} G$. If G and H are two finitely generated groups we may well want to say

$$\operatorname{Rel} H \leqslant_1 \operatorname{Rel} G,$$

and this makes sense. We shall not want to say

$$\text{`}\operatorname{Rel} H \subseteq \operatorname{Rel} G\text{'},$$

which would not make sense, even if $H \subseteq G$. Of course, for group elements $a_1, \dots, a_r$ and $b_1, \dots, b_r$ it does make sense, in fact important group-theoretic sense, to say

$$\operatorname{Rel}(b_1, \dots, b_r) \subseteq \operatorname{Rel}(a_1, \dots, a_r).$$

Next, we introduce the only recursion-theoretic reducibility that relativizes the notion of recursive enumerability rather than the notion of recursiveness.

Definition 4.7 For $X, Y \subseteq \mathbb{N}$, the set Y is said to be *enumeration reducible* to X, written $Y \leqslant_e X$, if there exists a recursively enumerable subset U of $\mathbb{N} \times \mathcal{P}_f(\mathbb{N})$ such that:

$$n \in Y \quad \text{if and only if} \quad (n, A) \in U, \qquad \text{for some } A \subseteq X.$$

We may emphasize the role of U by writing $Y \leqslant_e X$ via U.

Suppose that

$$X \leqslant_e Y \text{ via } U \quad \text{and} \quad Y \leqslant_e Z \text{ via } V,$$

where U and V are recursively enumerable subsets of $\mathbb{N} \times \mathcal{P}_f(\mathbb{N})$. We form a new set T as follows. Begin listing the elements of U and of V. Periodically, for each pair (n, A) that has so far appeared in the listing of U, check for each $m \in A$ whether or not there is a pair (m, B_m) in the listing of V. If there is such a pair for every $m \in A$, let

$$C = \bigcup \{B_m \mid m \in A\}$$

and print (n, C). (Form such a C for every possible combination of B_m, i.e. if (m, B_m) and (m, B'_m) both appear, form two sets C and C', one containing B_m and the other containing B'_{m}.) Let T be the set of all such (n, C), so

$$T = \{(n, C) \mid \exists A, \forall m \in A, \exists B_m((n, A) \in U \text{ and } (m, B_m) \in V)\}.$$

By construction and Church's Thesis, T is recursively enumerable, and it is easy to check that

$$X \leqslant_e Z \quad \text{via } T.$$

So $\leqslant_e$ is transitive.

Then $\leqslant_e$ induces a natural equivalence relation, written $\equiv_e$, on $\mathbb{N}$, with

$$Y \equiv_e X \quad \text{if and only if} \quad Y \leqslant_e X \text{ and } X \leqslant_e Y.$$

The corresponding equivalence classes form an upper semilattice, in which disjoint union is, as before, the least upper bound.

We note that if $f : \mathbb{N} \to \mathbb{N}$ is recursive, then

$$U = \{(n, \{f(n)\}) \mid n \in \mathbb{N}\} \subseteq \mathbb{N} \times \mathscr{P}_f(\mathbb{N})$$

is a recursively enumerable set, and

$$\text{if} \quad Y \leqslant_1 X \text{ via } f \quad \text{then} \quad Y \leqslant_e X \text{ via } U.$$

So $Y \leqslant_1 X$ implies that $Y \leqslant_e X$; but it is not hard to see that this implication does not go the other way, since there exists a set $Y \subseteq \mathbb{N}$ which is recursively enumerable but not empty, and then $Y \leqslant_e \varnothing$ but $Y \nleqslant_1 \varnothing$.

Each recursively enumerable subset U of $\mathbb{N} \times \mathscr{P}_f(\mathbb{N})$ determines a unique set Y such that $Y \leqslant_e X$ via U, for given $X \subseteq \mathbb{N}$. In fact,

$$Y = \{n \in \mathbb{N} \mid (n, A) \in U, \quad \text{for some} \quad A \subseteq X\}.$$

Thus, since there are only countably many recursively enumerable subsets of $\mathbb{N} \times \mathscr{P}_f(\mathbb{N})$, it follows that, for a given X, there are only countably many sets $Y \subseteq \mathbb{N}$ such that $Y \leqslant_e X$.

Note: $Y \leqslant_e \varnothing$ if and only if Y is recursively enumerable, and, if Y is recursively enumerable, then $Y \leqslant_e X$ for all $X \subseteq \mathbb{N}$. Also, if

$$Y_1 \leqslant_e X \text{ via } U \quad \text{and} \quad Y_2 \leqslant_e X \text{ via } V,$$

then

$$Y_1 \cup Y_2 \leqslant_e X \quad \text{via } U \cup V$$

and

$$Y_1 \cap Y_2 \leqslant_e X \quad \text{via } T,$$

where

$$T = \{(n, A \cup B) \mid (n, A) \in U, \quad (n, B) \in V\}.$$

Definition 4.8 Let $u, v_1, \ldots, v_k$ be words in the variables $x_1, \ldots, x_r$ and their inverses. then

$$(v_1 = v_2 = \cdots = v_k = 1) \;\Rightarrow\; (u = 1)$$

is called on *identical implication* if, in the free group freely generated by $x_1, \ldots, x_r$, the normal subgroup generated by $v_1, \ldots, v_k$ also contains u.

Lemma 4.9 *The set* U *of pairs* (u, A), *for*

$$u \in W(x_1, \ldots, x_r) = W \quad \text{and} \quad A = \{v_1, \ldots, v_k\} \subseteq W,$$

such that $(v_1 = \cdots = v_k = 1) \Rightarrow (u = 1)$ *is an identical implication, is recursively enumerable.*

Proof We have that $(u, A) \in U$ if and only if u is equal, in the free group freely generated by $x_1, \ldots, x_r$, to some word of the form

$$w_1^{-1} v_{i_1}^{\varepsilon_1} w_1 w_2^{-1} v_{i_2}^{\varepsilon_2} w_2 \cdots w_s^{-1} v_{i_s}^{\varepsilon_s} w_s,$$

where $w_j \in W$, $\varepsilon_j = \pm 1$, and $i_j \in \{1, \ldots, k\}$. We set a machine going, producing a list of all the words in $W(x_1, \ldots, x_r)$. Periodically we stop, at the nth place (say), and, if $u_1, \ldots, u_n$ are the words produced so far, we consider each

$$z = w_1^{-1} y_1^{\varepsilon_1} w_1 w_2^{-1} y_2^{\varepsilon_2} w_2 \cdots w_t^{-1} y_t^{\varepsilon_t} w_t,$$

for $\varepsilon_j = \pm 1$, w_j and y_j in $\{u_1, \ldots, u_n\}$, and $t \leqslant n$. Then we use the normal form theorem for free groups to decide whether $z = u_i$, for each $i \leqslant n$. Finally, for each $z = u_i$, we add $(u_i, \{y_1, \ldots, y_t\})$ to a list L. Clearly, $L = U$. □

Definition 4.10 Let $a_1, \ldots, a_r$ be elements of some group. Then a subset V of $\mathrm{Rel}\,(a_1, \ldots, a_r)$ is said to be a *set of defining relations* for $\langle a_1, \ldots, a_r \rangle$ if, for every $u \in \mathrm{Rel}\,(a_1, \ldots, a_r)$, there exist $v_1, \ldots, v_k$ in V such that

$$(v_1 = \cdots = v_k = 1) \quad \Rightarrow \quad (u = 1)$$

is an identical implication.

From Lemma 4.9, we see that the set

$$U = \{(w, \{z_1, \ldots, z_n\}) \mid (z_1 = \cdots = z_n = 1) \Rightarrow (w = 1) \text{ is an identical implication}\}$$

is a recursively enumerable set. From Definition 4.10, we see that $u \in \mathrm{Rel}\,(a_1, \ldots, a_r)$ if and only if $(u, \{v_1, \ldots, v_k\}) \in U$, for some $\{v_1, \ldots, v_k\} \subseteq V$. So we get the following result.

Theorem 4.11 *If* V *is a set of defining relations for* $G = \langle a_1, \ldots, a_r \rangle$ *then* $\mathrm{Rel}\,(a_1, \ldots, a_r) \leqslant_{\mathrm{e}} V$.

We now show that, for each $X \in \mathbb{N}$, there is an associated existentially closed group M_X such that, if $Y \subseteq \mathbb{N}$, then $M_X \cong M_Y$ if and only if $X \equiv_{\mathrm{e}} Y$.

Theorem 4.12 *For any subset X of $\mathbb{N}$, there exists a finitely generated group G such that $X \equiv_e \text{Rel}\, G$.*

Proof Take distinct elements $a, b, c,$ and d. Let $b_i = a^{-i}ba^i$ and $d_i = c^{-i}dc^i$, for $i \in \mathbb{N}$. Then let

$$G = \langle a, b, c, d \mid b_j = d_j, \quad \forall j \in X \rangle;$$

so G is the free product of the free groups $\langle a, b \rangle$ and $\langle c, d \rangle$, amalgamating the groups $\langle b_j \mid j \in X \rangle$ and $\langle d_j \mid j \in X \rangle$. By the normal form theorem for free products, $b_i d_i^{-1} \in \text{Rel}\,(a, b, c, d)$ if and only if $i \in X$, and so $X \leqslant_1 \text{Rel}\,(a, b, c, d)$. We may take $V = \{b_i d_i^{-1} \mid i \in X\}$ as a set of defining relations for G, and so, by Theorem 4.11, $\text{Rel}\, G \leqslant_e X$, giving the result. □

Notice that it cannot be true that, for every $X \subseteq \mathbb{N}$, there exists a finitely generated group G such that $X \equiv_1 \text{Rel}\, G$, since $\text{Rel}\, G$ cannot be finite.

Definition 4.13 An *ideal* J is a non-empty subset of an upper semilattice L such that:

(i) If $A \in J$ and $B \in L$ with $B \leqslant A$, then $B \in J$.

(ii) *If* $A, B \in J$ then the least upper bound $A \vee B$ of A and B lies in J.

The set $\mathcal{P}(\mathbb{N})/\equiv_e$, of equivalence classes of $\mathcal{P}(\mathbb{N})$ under enumeration reducibility, is an upper semilattice. For any subset X of $\mathbb{N}$, the set $\mathrm{J}(X)$ of all equivalence classes that are generated by some Y such that $Y \leqslant_e X$, is a countable (by one of the remarks above Definition 4.8) ideal of $\mathcal{P}(\mathbb{N})/\equiv_e$, and $\mathrm{J}(X)$ is called the *principal ideal* of $\mathcal{P}(\mathbb{N})/\equiv_e$ generated by X. There are also countable ideals of $\mathcal{P}(\mathbb{N})/\equiv_e$ which are not of the form $\mathrm{J}(X)$, for any $X \subseteq \mathbb{N}$. For example, for each $Y \subseteq \mathbb{N}$, there exists (since $\mathcal{P}(\mathbb{N})$ is uncountable) some $Z \subseteq \mathbb{N}$ such that $Z \not\leqslant_e Y$. So $Y \leqslant_e Z \vee Y$ but $Z \vee Y \not\leqslant_e Y$. Thus, we can find a sequence $X_0, X_1, X_2, \ldots$ of elements of $\mathcal{P}(\mathbb{N})$ such that $X_i \leqslant_e X_{i+1}$, but $X_{i+1} \not\leqslant_e X_i$, for all $i \in \mathbb{N}$. Then the set J of all equivalence classes generated by the sets Y such that $Y \leqslant_e X_i$, for some i, is a countable ideal. But, given any X which generates an equivalence class in J, there exists $i \in \mathbb{N}$ such that $X \leqslant_e X_i$, and so $X_{i+1} \not\leqslant_e X$. Thus $J \neq \mathrm{J}(X)$, for any $X \subseteq \mathbb{N}$.

Theorem 4.14 *For any ideal J of $\mathcal{P}(\mathbb{N})/\equiv_e$, let $\mathcal{G}(J)$ be the class of finitely generated groups G such that the equivalence class containing* $\text{Rel}\, G$ *lies in J. Then $\mathcal{G}(J)$ satisfies* SC, JEP, *and* AC (*as defined in Chapter* 3).

Proof By Theorem 4.5 and the definition of an ideal, $\mathcal{G}(J)$ satisfies SC.

For finitely generated groups $G = \langle g_1, \ldots, g_n \rangle$ and $H = \langle h_1, \ldots, h_m \rangle$, by the normal form theorem for free products, we can take $\mathrm{Rel}\,(g_1, \dots, g_n) \vee \mathrm{Rel}\,(h_1, \dots, h_m)$ as a set of defining relations for the free product $G * H$. So

$$\mathrm{Rel}\,(G * H) \leqslant_{\mathrm{e}} \mathrm{Rel}\, G \vee \mathrm{Rel}\, H,$$

and then, if $G, H \in \mathscr{G}(J)$, we have $G * H \in \mathscr{G}(J)$, and so $\mathscr{G}(J)$ satisfies JEP.

If $G \in \mathscr{G}(J)$ and $\mathscr{S}$ is a finite set of equations defined over G and soluble over G, then there is a group K, containing G, in which $\mathscr{S}$ is soluble. Let $h_1, \dots, h_s$ be solutions of $\mathscr{S}$ in K. Let

$$H = \langle G, h_1, \dots, h_s \mid \mathscr{S}, \mathrm{Rel}\, G \rangle,$$

then, as in Theorem 3.10, $G \subseteq H$, H is finitely generated, $\mathscr{S}$ is soluble in H and, as defining relations for H, we can take $\mathrm{Rel}\, G \vee \mathscr{S}$. Since $\mathscr{S}$ is finite, we have $\mathrm{Rel}\, H \leqslant_{\mathrm{e}} \mathrm{Rel}\, G \vee \mathscr{S} \equiv_{\mathrm{e}} \mathrm{Rel}\, G$. So $H \in \mathscr{G}(J)$, and thus $\mathscr{G}(J)$ satisfies AC. □

From Theorem 4.14, Theorem 3.9, and Theorem 3.4 we get the following result.

Theorem 4.15 *If J is a countable ideal of $\mathscr{P}(\mathbb{N})/\equiv_{\mathrm{e}}$, then there exists one and (up to isomorphism) only one countable existentially closed group M_J such that*

$$\mathscr{G}(J) = \mathrm{Sk}\, \mathrm{M}_J.$$

Proof If J is a countable ideal, let $X_0, X_1, \dots$ be representatives of the equivalence classes in J. Then, for each $G \in \mathscr{G}(J)$, there exists $i \in \mathbb{N}$ such that $\mathrm{Rel}\, G \equiv_{\mathrm{e}} X_i$. Since G is finitely generated, G is isomorphic to some quotient F/R of some free group F of finite rank. There are only countably many such free groups, up to isomorphism, and there are only countably many such sets $Y \subseteq \mathbb{N}$ such that $Y \equiv_{\mathrm{e}} X_i$. So there are, up to isomorphism, only countably many quotient groups F/R with $\mathrm{Rel}\,(F/R) \equiv_{\mathrm{e}} X_i$. So $\mathscr{G}(J)$ contains only countably many isomorphism types of groups. By definition, J is non-empty; so, by Theorem 4.12, $\mathscr{G}(J)$ is non-empty. Since all finite groups have recursively enumerable relation sets, $\mathrm{Rel}\,\{1\} \equiv_{\mathrm{e}} \mathrm{Rel}\, H$, for all finite groups H. Thus, if $\mathscr{G}(J)$ contains the trivial group, it contains all finite groups and hence is a non-trivial class. We can thus use Theorem 4.14, Theorem 3.9, and Theorem 3.4, to get the result. □

Definition 4.16 In the case when J is the principal ideal generated by $X \subseteq \mathbb{N}$, we write M_X instead of M_J or $\mathrm{M}_{J(X)}$; thus

$$\mathrm{Sk}(\mathrm{M}_X) = \{G \mid G \text{ is finitely generated and } \mathrm{Rel}\, G \leqslant_{\mathrm{e}} X\}.$$

Remarks

(i) Suppose that $\mathrm{M}_X \cong \mathrm{M}_Y$. By Theorem 4.12, we can find a finitely generated group G such that $\mathrm{Rel}\, G \equiv_e X$. So $G \in \mathrm{Sk}\, \mathrm{M}_X = \mathrm{Sk}\, \mathrm{M}_Y$ and hence $\mathrm{Rel}\, G \leqslant_e Y$. Similarly, $Y \leqslant_e X$ and so $Y \equiv_e X$. Thus $\mathrm{M}_X \cong \mathrm{M}_Y$ if and only if $Y \equiv_e X$.

Since each $\equiv_e$-equivalence class is countable and $\mathscr{P}(\mathbb{N})$ has cardinality $2^{\aleph_0}$, from Theorem 4.15 we see that there are $2^{\aleph_0}$ non-isomorphic countable existentially closed groups.

(ii) Each countable group can be embedded in a finitely generated group H, and thus in a countable existentially closed group M_X, where $X = \mathrm{Rel}\, H$.

(iii) In contrast to the general case (see Theorem 8.1), if M_X is a countable existentially closed group of the type in Definition 4.16, with $G, H \in \mathrm{Sk}\, \mathrm{M}_X$ and $U = G \cap H$ finitely generated, then the free product $G *_U H$ is an element of $\mathrm{Sk}\, \mathrm{M}_X$. As defining relations for $G *_U H$, we can take the relations of G and H, together with a finite set of relations identifying the elements of U in G, with the elements of U in H. Thus,

$$\mathrm{Rel}\,(G *_U H) \equiv_e (\mathrm{Rel}\, G) \vee (\mathrm{Rel}\, H) \leqslant_e X,$$

and this is sufficient to guarantee that $G *_U H \in \mathrm{Sk}\, \mathrm{M}_X$.

(iv) For any existentially closed group M, the direct product $M \times M$ lies in M. For any $X \subseteq \mathbb{N}$, we have $\mathrm{M}_X * \mathrm{M}_X \subseteq \mathrm{M}_X$. It is easy to see, using The Subgroup Theorem for finitely presented groups (Higman, 1961) that $\mathrm{M}_\varnothing$ is not closed under taking wreath products. It is also possible, but much harder, to see that no group of the form M_X is closed under the taking of wreath products.

(v) Each group M_X has a finitely generated subgroup which is relatively universal, i.e. there exists $G \in \mathrm{Sk}\, \mathrm{M}_X$ such that $\mathrm{Sk}\, \mathrm{M}_X = \mathrm{Sk}\, G$ (see Corollary 6.4).

Next we want to prove some results, about recursive and recursively enumerable sets, that we shall need later. In order to do this, we need some discussion of partial recursive functions and their effective enumeration.

A standard diagonalization argument shows that the set of all recursive functions is not recursively enumerable. If $Z = \{f_i \mid i \in \mathbb{N}\}$ is some recursively enumerated set of recursive functions, the function g defined by $g(n) = f_n(n) + 1$, $\forall\, n \in \mathbb{N}$, is recursive, but clearly $g \notin Z$. Results in the area that we now consider derive primarily from the tension between the fact that machines can be enumerated but recursive functions cannot. One obvious reason for this difference is that, while a recursive function is always determined by a machine, we cannot always tell whether or not a machine determines a function (although if it does, the function will be recursive).

For example, it is easy to see that there exist machines N_1, N_2, and N_3, each of which accepts any natural number $n \in \mathbb{N}$ as input, and which have the following properties, respectively.

(1) For input $n \in \mathbb{N}$, machine N_1 produces as output the smallest prime p such that $p \geqslant n$.

(2) For input $n \in \mathbb{N}$, machine N_2 produces as output the smallest prime $p \geqslant n$ such that $p+1$ is also prime, if such a p exists; otherwise N_2 produces no output for this n.

(3) For input $n \in \mathbb{N}$, machine N_3 produces as output the smallest prime $p \geqslant n$ such that $p+2$ is also prime, if such a p exists; otherwise N_3 produces no output for this n.

Then N_1 defines a function, but N_2 does not define a function since it gives no output for $n \geqslant 3$, and N_3 may or may not define a function, depending on the truth of the twin-prime conjecture.

Essentially, our arguments about recursive and recursively enumerable sets will be based on the assumption that this doubt, as to whether a machine will produce an output or not for a given input, is the only difference between machines and functions.

More precisely, with each machine M, we can associate a partial function f such that if, for input n, the machine M produces output m, then $f(n)=m$, but if, for input n, the machine produces no output, then $f(n)$ is undefined. A partial function which can be associated with a machine in this way is called a *partial recursive function* (or recursive partial function). We note that a partial recursive function is a total function if and only if it is a recursive function. Of course, two different machines may determine the same partial function, since we can always introduce trivial steps into one machine to get another machine. Since the set of machines can be effectively enumerated, there is a corresponding effective enumeration of the set of partial recursive functions. The enumeration of the partial recursive functions may (in fact will have to) contain repetitions. Recall that we are taking the definition of machine to imply that the set of all machines can be mechanically enumerated. So we can define a partial function f, of two variables, by the rule that $f(i, n)$ is the result of giving the ith machine the input n, and f will be recursive.

Any partial recursive function is defined by some machine; so, given any partial recursive function g of one variable, there exists some $i \in \mathbb{N}$ such that

$$g(n)=f(i, n) \quad \forall n \in \mathbb{N},$$

(in the sense that $g(n)$ and $f(i, n)$ are both defined and equal, or are both undefined). The integer i associated with g corresponds, in some effective enumeration of machines, to a machine which calculates g. As we have already remarked, we cannot make the choice of i unique.

If we write $f_i(n)$ for $f(i, n)$, we get an effectively enumerated sequence

$$f_0, f_1, f_2, \ldots .$$

of all partial recursive functions of one variable, corresponding to a given enumeration

$$M_0, M_1, M_2, \ldots$$

of all machines. This automatically gives us a corresponding effective enumeration

$$E_0, E_1, E_2, \ldots$$

of all the recursively enumerable subsets of $\mathbb{N}$, where

$$E_i = \{m \mid \exists n(f_i(n) = m)\}.$$

The fact that the indices in the above enumeration of the partial recursive functions correspond to machines, enables us to make a useful application of Church's Thesis. If we have a mechanical process which we can apply to each machine M_i to get a new machine $M_{p(i)}$, Church's Thesis tells us that the function p will be recursive. To calculate $p(i)$, we apply the process to M_i to produce $M_{p(i)}$, and then, since machines are all finite, we can check $M_1, M_2, \ldots$ until we find j such that

$$M_{p(i)} = M_j.$$

Thus $p(i) = j$. So we can calculate p mechanically; hence, by Church's Thesis, p is recursive.

For instance, there is a recursive function p such that

$$f_{p(i)}(n) = f_i(n) + 1,$$

since we can obtain $M_{p(i)}$ from M_i by adding a new last line

'add 1'

to M_i, before the output is produced.

In a similar way, if there is a mechanical process which can be applied to pairs (M_i, M_j) of machines to yield a new machine $M_{q(i,j)}$, then q is also a recursive function, now of two variables. For instance, we could take $M_{q(i,j)}$ to include both M_i and M_j, in such a way that for given input n, the machine $M_{q(i,j)}$ sets M_i going with input n, and M_j going with successive inputs $0, 1, 2, \ldots$. Then, if M_i produces output $f_i(n)$, and if, for some $k \in \mathbb{N}$, the machine M_j produces output $f_j(k)$ such that

$$f_i(n) = f_j(k),$$

then $M_{q(i,j)}$ produces output $f_i(n)$. So we see that there is a recursive function q, of two variables, such that $f_{q(i,j)}(n)$ is defined if and only if

$f_i(n)$ is defined and equal to $f_j(k)$, for some $k \in \mathbb{N}$. Further, in this case,

$$f_{q(i,j)}(n) = f_i(n).$$

Thus,

$$E_{q(i,j)} = E_i \cap E_j.$$

Lemma 4.17 *There exists a recursively enumerable set X such that $\mathbb{N} \backslash X$ is not recursively enumerable.*

Proof Let

$$f_0, f_1, f_2, \ldots$$

be an effective enumeration of the partial recursive functions, and let f be defined by

$$f(i, n) = f_i(n).$$

Since f is a partial recursive function of two variables, the set

$$U = \{(i, j) \mid \exists n(f_i(n) = j)\}$$

is recursively enumerable. Now,

$$U = \{(i, j) \mid j \in E_i\}$$

and so the set

$$X = \{i \mid i \in E_i\}$$

is recursively enumerable. If $\mathbb{N} \backslash X$ is recursively enumerable, then $\mathbb{N} \backslash X = E_j$, for some $j \in \mathbb{N}$. But $j \in \mathbb{N} \backslash X$ if and only if $j \notin E_j$, which is a contradiction. So $\mathbb{N} \backslash X$ is not recursively enumerable, and X is the required set. □

Lemma 4.18 *For any subset A of $\mathbb{N}$ there exists a subset $X \subseteq \mathbb{N}$ such that $X \leqslant_e A$ but $\mathbb{N} \backslash X \not\leqslant_e A$.*

Proof There is an effective enumeration $U_0, U_1, U_2, \ldots$ of the recursively enumerable subsets of $\mathbb{N} \times \mathcal{P}_f(\mathbb{N})$. If $Z \leqslant_e A$, then $Z \leqslant_e A$ via U_i, for some i. We denote by $U_i(A)$ the unique subset of $\mathbb{N}$ which is enumeration-reducible to A via U_i. So we have a sequence $U_0(A)$, $U_1(A)$, $U_2(A), \ldots$ of subsets of $\mathbb{N}$ such that $Z \leqslant_e A$ if and only if $Z = U_i(A)$, for some $i \in \mathbb{N}$. Further, the set $P = \{(i, j, C) \mid (i, C) \in U_j\}$ is recursively enumerable, and so

$$Y = \{(i, j) \mid i \in U_j(A)\} \leqslant_e A \quad \text{via } P.$$

Let

$$X = \{i \in \mathbb{N} \mid i \in U_i(A)\}.$$

Then, since

$$\{(i, \{(i, i)\}) \mid i \in \mathbb{N}\}$$

is recursively enumerable, $X \leqslant_e Y \leqslant_e A$. But, if $\mathbb{N} \backslash X \leqslant_e A$, then $\mathbb{N} \backslash X = U_j(A)$ for some $j \in \mathbb{N}$; so, as for the previous lemma, $\mathbb{N} \backslash X \nleqslant_e A$. □

The next two results fit naturally into a discussion of recursive functions, but they will not be used until Chapter 5.

Definition 4.19 A recursively enumerable set $X \subseteq \mathbb{N}$ is said to be *strongly creative,* if there exists a (total) injective recursive function g such that $g(i)$ belongs either to $X \cap E_i$ or $\mathbb{N} \backslash (X \cup E_i)$, for all i (where E_i is the ith recursively enumerable set, as above).

Note that, if X is a recursive set, then $\mathbb{N} \backslash X$ is recursively enumerable, so $\mathbb{N} \backslash X = E_i$, for some i. But $X \cap (\mathbb{N} \backslash X) = \mathbb{N} \backslash (X \cup (\mathbb{N} \backslash X)) = \varnothing$, so X cannot be strongly creative.

Our definition of 'strongly creative' is a variant of the standard definition, which seems to fit our purposes.

Theorem 4.20 *There exist recursively enumerable sets X and Y such that every recursively enumerable set Z, with* $X \subseteq Z$ *and* $Y \cap Z = \varnothing$, *is strongly creative.*

Proof Take $X = \{i \mid f_i(i) = 0\}$ and take $Y = \{i \mid f_i(i) = 1\}$. Let Z be any recursively enumerable set such that $X \subseteq Z$ and $Y \cap Z = \varnothing$. Let N and M_j be machines which enumerate Z and E_j, respectively. We construct a new machine to calculate a partial function g. Set N and M_j going and put $g(n) = 1$, if n is produced first by N, and $g(n) = 0$, if n is produced first by M_j. (If n is produced at the same time by both M_j and N, we put $g(n) = 0$.) From the above discussion, we see that there exists a recursive function h, which will be injective, such that $g = f_{h(j)}$. We consider the functions $f_{h(i)}$, for each $i \in \mathbb{N}$.

If $f_{h(i)}(h(i))$ is not defined, then $h(i) \notin Z \cup E_i$, so $h(i) \notin X \subseteq Z$ and $h(i) \in \mathbb{N} \backslash (X \cup E_i)$.

If $f_{h(i)}(h(i)) = 0$ then, by definition, $h(i) \in X \subseteq Z$. But, also by definition of $f_{h(i)}$, we have that $h(i)$ must have been produced by M_i, so $h(i) \in X \cap E_i \subseteq Z \cap E_i$.

If $f_{h(i)}(h(i)) = 1$ then $h(i) \in Y$ and so $h(i) \notin Z$. But $h(i)$ must also have been produced by N, and so $h(i) \in Z$. This is a contradiction, so this case cannot arise.

Thus $h(i) \in Z \cap E_i$ or $h(i) \in \mathbb{N} \backslash (Z \cup E_i)$, as required. □

Definition 4.21 A set $X \subseteq \mathbb{N}$ is said to be *1-complete* if it is recursively enumerable and if $Y \leqslant_1 X$ for every recursively enumerable set Y.

Lemma 4.22 *A strongly creative set is 1-complete.*

Proof Let X be a strongly creative set with creative function g. Let Y be a recursively enumerable set; let

$$M_0, M_1, M_2, \ldots$$

be an enumeration of all machines; and let

$$E_0, E_1, E_2, \ldots$$

be the corresponding recursively enumerable sets. Let M be the machine which enumerates Y.

For each $i \geq 1$, we obtain a new machine N_i from M_i as follows. For input n, machine N_i sets M going and then produces output n if and only if, at some stage, M produces output i. So we obtain N_i from M_i by deleting everything in M_i and then adding M, together with an extra step:

'if M produces i, print n.'

Let f_i denote the partial function defined by N_i. Then f_i is either the (total) identity function on $\mathbb{N}$, or else f_i is nowhere defined.

By Church's Thesis, we can find a recursive function p such that $N_i = M_{p(i)}$. By adding a different trivial step to each N_i, we may assume that p is injective.

Now, $E_{p(i)}$ is the range of f_i, so $E_{p(i)} = \mathbb{N}$ or $E_{p(i)} = \varnothing$. By definition of g,

$$g(p(i)) \in (X \cap E_{p(i)}) \cup (\mathbb{N} \setminus (X \cup E_{p(i)})).$$

Since either $i \in Y$ and $E_{p(i)} = \mathbb{N}$, or $i \notin Y$ and $E_{p(i)} = \varnothing$, we have that $g(p(i)) \in X$ if and only if $i \in Y$. Thus,

$$Y \leq_1 X \quad \text{via} \quad g \circ p,$$

as required. □

We now return to recursion-theoretic reducibilities, beginning with two that play only a minor role in this work.

Definitions For subsets X and Y of $\mathbb{N}$ we say that:

(a) Y is *Turing reducible* to X, written $Y \leq_T X$, if there exist recursively enumerable sets U and V in $\mathbb{N} \times \mathcal{P}_f(\mathbb{N}) \times \mathcal{P}_f(\mathbb{N})$ such that

(i) $n \in Y \Leftrightarrow (n, A, B) \in U$, for some $A \subseteq X$ and some $B \subseteq \mathbb{N} \setminus X$,

(ii) $n \notin Y \Leftrightarrow (n, A, B) \in V$, for some $A \subseteq X$ and some $B \subseteq \mathbb{N} \setminus X$.

(b) Y is *positively reducible* to X, written $Y \leq^+ X$, if there is a recursive function $f: \mathbb{N} \to \mathcal{P}_f(\mathbb{N}) \cup \{\infty\}$, such that $n \in Y$ if and only if $f(n) \neq \infty$ and $f(n) \subseteq X$.

Notice that $Y \leq_1 X$ implies that $Y \leq^+ X$, and that $Y \leq^+ X$ implies that

$Y \leqslant_T X$ and that $Y \leqslant_e X$. But there are no valid implications between $Y \leqslant_e X$ and $Y \leqslant_T X$.

Next we define a group G_X, for each $X \subseteq \mathbb{N}$. This group is generated by two elements a and b subject to the defining relations

$$\{ac_i = c_i a, \quad bc_i = c_i b \mid i \in \mathbb{N}\} \cup \{c_i^2 = 1 \mid i \in \mathbb{N}\} \cup \{c_j = 1 \mid j \in X\},$$

where $b_i = a^{-i}ba^i$ and $c_i = [b, b_{i+1}]$. The elements c_i belong to the centre of G_X and so $c_{-i-2} = c_i^{-1} = c_i$, and $b_i^{-1}b_j^{-1}b_i b_j = c_{j-i-1}$, for all i,j in $\mathbb{N}$. The b_i generate the subgroup of G_X that consists precisely of those words in a and b in which the sum of the powers of a is zero. Thus, every element of G_X can be written in the form

$$a^{\alpha} b_{j_1}^{\beta_1} \cdots b_{j_k}^{\beta_k} c_{l_1} \cdots c_{l_h},$$

where $\alpha, j_s, \beta_s, l_t \in \mathbb{Z}$ $(1 \leqslant s \leqslant k,\ 1 \leqslant t \leqslant h)$, $\beta_s \neq 0$, $j_1 < \ldots < j_k$, and $0 \leqslant l_1 < l_2 < \ldots < l_h$. (We may have h and k equal to zero.)
We call a and b the *standard generators* for G_X.

We note that G_X is a homomorphic image of a centre-by-metabelian group, which was constructed by P. Hall to exhibit the failure of certain chain conditions. By a result originally due to Hall, we have that

$$a^{\alpha} b_{j_1}^{\beta_1} \cdots b_{j_k}^{\beta_k} c_{l_1} \cdots c_{l_h} = 1$$

if and only if $\alpha = k = 0$, and $l_i \in X$, for all $i \in \{1, \ldots, n\}$. Thus the above expression gives us a normal form for the elements of $G_\varnothing$, and the map from G_X to $G_\varnothing$ which takes an element to its associated normal form is clearly recursive.

Theorem 4.23 *For any* $X \subseteq \mathbb{N}$,

$$X \leqslant_1 \operatorname{Rel} G_X \leqslant^+ X.$$

Proof Choose variables x and y, and let $W(x, y)$ denote the set of all words on x, y, x^{-1}, and y^{-1}. Define $y_i = x^{-i}yx^i$ and $z_i = y^{-1}y_i^{-1}yy_i$. The map $i \to z_i$ is a recursive injective map from $\mathbb{N}$ to $W(x, y)$, and $i \in X$ if and only if $z_i \in \operatorname{Rel}(a, b)$, where a and b are standard generators for G_X. So $X \leqslant_1 \operatorname{Rel} G_X$.

We construct a map $f: W(x, y) \to \mathcal{P}_f(\mathbb{N}) \cup \{\infty\}$ using the normal form. For $w \in W(x, y)$, suppose that $w(a, b) = a^{\alpha} b_{j_1}^{\beta_1} \cdots c_{l_h}$. Then we set $f(w) = \infty$, if $\alpha \neq 0$ or if $k \geqslant 1$, and set $f(w) = \{l_1, \ldots, l_h\}$ otherwise. Then $w \in \operatorname{Rel}(a, b)$ if and only if $f(w) \neq \infty$ and $f(w) \subseteq X$, giving the result. □

Corollary 4.24 *For any subset* $X \subseteq \mathbb{N}$ *there exists a finitely generated group* G *such that* $\operatorname{Rel} G \equiv^+ X$.

Theorem 4.25 *There exist finitely generated groups* G *and* H, *and an existentially closed group* M, *such that* $\operatorname{Rel} G \equiv_T \operatorname{Rel} H$ *and* $G \in \operatorname{Sk} M$ *but* $H \notin \operatorname{Sk} M$.

Proof Putting $U' = \{(n, B, A) \mid (n, A, B) \in V\}$ and

$$V' = \{(n, B, A) \mid (n, A, B) \in U\}$$

in the definition of $\leqslant_T$, we see that $Y \leqslant_T X$ if and only if $\mathbb{N}\backslash Y \leqslant_T \mathbb{N}\backslash X$. Putting $U_1 = \{(n, \varnothing, \{n\}) \mid n \in \mathbb{N}\}$ and $V_1 = \{(n, \{n\}, \varnothing) \mid n \in \mathbb{N}\}$, we see that $\mathbb{N}\backslash X \leqslant_T X$. Thus, from Theorem 4.23,

$$\text{Rel } \mathrm{G}_X \equiv_T X \equiv_T \mathbb{N}\backslash X \equiv_T \text{Rel } \mathrm{G}_{\mathbb{N}\backslash X}.$$

Take X, as in Lemma 4.17, to be a recursively enumerable set such that $\mathbb{N}\backslash X$ is not recursively enumerable. Then $\text{Rel } \mathrm{G}_X \equiv_e X$, and so $\text{Rel } \mathrm{G}_X$ is recursively enumerable, but $\text{Rel } \mathrm{G}_{\mathbb{N}\backslash X} \equiv_e \mathbb{N}\backslash X$, and so $\text{Rel } \mathrm{G}_{\mathbb{N}\backslash X}$ is not recursively enumerable. Putting $G = \mathrm{G}_X$, $H = \mathrm{G}_{\mathbb{N}\backslash X}$, and $M = \mathrm{M}_{\varnothing}$, we get the result. □

The main reason for introducing into this text yet another reducibility is to avoid the analogue of the behaviour at Turing reducibility that is described in Theorem 4.25. The new equivalence $\leqslant^*$ has the property that, for any finitely generated groups G and H, every existentially closed group that embeds G also embeds H if and only if $\text{Rel } H \leqslant^* \text{Rel } G$. However, we will not prove this result until much later.

Definitions Let I, J, X, Y be subsets of $\mathbb{N}$. We say that:

(a) The *pair* (Y, J) is *Ziegler reducible* to the pair (X, I), written $(Y, J) \leqslant^* (X, I)$, if $Y \leqslant_e X$ and if there exists a recursively enumerable subset U of $\mathbb{N} \times \mathscr{P}_f(\mathbb{N}) \times (\mathbb{N} \cup \{\infty\})$, such that $n \in J$ if and only if there exists some $(n, A, m) \in U$, with $A \subseteq X$ and $m \in I \cup \{\infty\}$.

(b) The *set* Y is *Ziegler reducible* to X, written $Y \leqslant^* X$, if $(Y, \mathbb{N}\backslash Y) \leqslant^* (X, \mathbb{N}\backslash X)$ in the sense of (a).

Notice that $Y \leqslant^* X$ if and only if

(i) $Y \leqslant_e X$,

(ii) there exists a recursively enumerable subset U of

$$\mathbb{N} \times \mathscr{P}_f(\mathbb{N}) \times (\mathbb{N} \cup \{\infty\}),$$

such that $n \notin Y$ if and only if $(n, A, m) \in U$ for some $A \subseteq X$ and for some $m \notin X$.

Also notice that $Y \leqslant^+ X$ implies $Y \leqslant^* X$, which in turn implies $Y \leqslant_T X$.

Because it is mainly Ziegler reducibility that we are concerned with, we put on record the following version of Corollary 4.24.

Theorem 4.26 *For any $X \subseteq \mathbb{N}$, the finitely generated group G_X has the property that* $\text{Rel } \mathrm{G}_X \equiv^* X$.

Let $\{x_0, x_1, x_2, \dots\}$ be a set of variables, and let F be the free group freely generated by these variables. We take sets U and V of words on the x_i and their inverses, and we say that the pair (U, V) is *consistent* if there exists some group G in which the equations $\{u=1 \mid u \in U\}$ and the inequalities $\{v \neq 1 \mid v \in V\}$ are solvable. For any subset Y of $\mathrm{W}(\boldsymbol{x})$ (the set of all words on $x_0^{\pm 1}, x_1^{\pm 1}, x_2^{\pm 1}, \ldots$), write $\langle Y \rangle^F$ for the smallest subset X of $\mathrm{W}(\boldsymbol{x})$ that contains Y and is closed under the rules

(i) $w \in X$ implies that $w^{-1} \in X$ (w^{-1} = the formal universe of w),
(ii) $y \in X$ and $w \in \mathrm{W}$ imply that $w^{-1}yw \in X$,
(iii) w and $y \in X$ imply that $wy \in X$.

So, under the natural correspondence $\mathrm{W}(\boldsymbol{x}) \to F$, the set $\langle Y \rangle^F$ corresponds to the normal subgroup of F generated by the image of Y in F. We say that $w=1$ is a *consequence* of U, and of (U, V), if $w \in \langle U \rangle^F$, and that $z \neq 1$ is a *consequence of the pair* (U, V) if $\langle U, z \rangle^F \cap V \neq \varnothing$. Thus, $w=1$ and $z \neq 1$ hold in any group in which $U=\{1\}$ and $V \neq \{1\}$ hold. We denote

$$\bar{U} = \langle U \rangle^F \quad \text{and} \quad \bar{V}_U = \{z \mid \langle U, z \rangle^F \cap V \neq \varnothing\}.$$

Theorem 4.27 *If the pair* (U, V) *is consistent, then* $(\bar{U}, \bar{V}_U) \leqslant^* (U, V)$.

Proof There are only countably many finite subsets of a countable set. For each $A \subseteq \mathrm{W}(\boldsymbol{x})$ we can recursively enumerate the set $\{w=1 \mid w=\bar{A}\}$ of all consequences of A. Hence the set

$$Z = \{(w, A) \mid A \text{ finite}, \quad w \in \bar{A}\}$$

is recursively enumerable. Any consequence of U is a consequence of some finite subset of U, and so

$$\bar{U} \leqslant_{\mathrm{e}} U \quad \text{via} \quad Z.$$

Now, if F is the free group generated by $x_0, x_1, \ldots$, then $z \in \bar{V}_U$ if and only if $\langle A, z \rangle^F$ contains an element of V, for some finite set $A \subseteq U$. The set

$$Y = \{(z, A, v) \mid A \subseteq \mathscr{P}_{\mathrm{f}}(\mathrm{W}(\boldsymbol{x})), \quad z \in \mathrm{W}(\boldsymbol{x}), \quad v \in \langle A, z \rangle^F\}$$

is certainly recursively enumerable, and $z \in \bar{V}_U$ if and only if $v \in \langle A, z \rangle^F$, for some $A \subseteq U$ and some $v \in V$. So

$$(\bar{U}, \bar{V}_U) \leqslant^* (U, V),$$

as required. □

We note the following results, which we leave to the reader to prove.

Lemma 4.28

(i) *For* $U, V, W, X, Y, Z \subseteq \mathbb{N}$, if $(U, V) \leqslant^* (X, Y)$ and $(W, Z) \leqslant^* (X, Y)$ then

$$(W \cup U, \quad Z \cup V) \leqslant^* (X, Y).$$

(ii) *For any $X, Y \subseteq \mathbb{N}$ and any recursively enumerable set $R \subseteq \mathbb{N}$,*

$$(X \cap R, \quad Y \cap R) \leqslant^* (X, Y).$$

Theorem 4.29 *Let (U, V) be a consistent pair of sets of words, on the variables $x_0, x_1, \dots, x_r$ and their inverses. Let $\{y_0, y_1, \dots, y_s\}$ be a second set of variables and let $\{z_0, z_1, \dots, z_s\}$ be a set of words in $W(\boldsymbol{x})$. If*

$$X = \{w(y_0, \dots, y_s) \mid w(z_0, \dots, z_s) \in \bar{U}\},$$

$$Y = \{w(y_0, \dots, y_s) \mid w(z_0, \dots, z_s) \in \bar{V}_U\},$$

then $(X, Y) \leqslant^* (U, V)$.

Proof The set S of all words of the form $w(z_0, \dots, z_s)$ is a recursively enumerable subset of $\mathrm{W}(x_1, \dots, x_r)$. Since $X = \{w(\boldsymbol{y}) \mid w(\boldsymbol{z}) \in S \cap \bar{U}\}$ and $Y = \{w(\boldsymbol{y}) \mid w(\boldsymbol{z}) \in S \cap \bar{V}_U\}$ and since the sets

$$\{(w(\boldsymbol{y}), \ \{w(\boldsymbol{z})\}) \mid w \in S\}, \qquad \{(w(\boldsymbol{z}), \ \{w(\boldsymbol{y})\}) \mid w \in S\},$$

$$\{(w(\boldsymbol{y}), \ \varnothing, \ w(\boldsymbol{z})) \mid w \in S\}, \qquad \{(w(\boldsymbol{z}), \ \varnothing, \ w(\boldsymbol{y})) \mid w \in S\}$$

are all recursively enumerable, we get that $X \equiv_e S \cap \bar{U}$ and $(X, Y) \equiv^* (S \cap \bar{U}, \ S \cap \bar{V}_U)$. Then, by Theorem 4.27 and Lemma 4.28, we have $(X, Y) \leqslant^* (U, V)$, as required.

Corollary 4.30 *Let G and H be finitely generated groups such that $H \subseteq G$*; then $\mathrm{Rel}\, H \leqslant^* \mathrm{Rel}\, G$.

5
Applications of The Subgroup Theorem

In this chapter we consider applications of the following theorem (Higman, 1961).

The Subgroup Theorem *A finitely generated group G can be embedded in a finitely presented group if and only if* Rel G *is recursively enumerable.*

We define what we mean by positive and negative implications over G, and we define the group $\langle G, \mathscr{S} \rangle$, the group obtained by freely adjoining to G the solutions of the set $\mathscr{S}$ of equations and positive implications over G. We then apply The Subgroup Theorem to prove that if M is any existentially closed group, then M embeds every finitely generated group with soluble word problem, M embeds some finitely generated group with unsolvable word problem, and, if $G \in \operatorname{Sk} M$ and $\operatorname{Rel} H \leqslant^* \operatorname{Rel} G$, then $H \in \operatorname{Sk} M$. (A finitely generated group G has solvable word problem if Rel G is recursive. So if $\{a_1, \ldots . a_r\}$ is a generating set for G, then G has solvable word problem if and only if $\operatorname{Rel}(a_1, \ldots, a_r)$ is recursive.)

However, we begin by proving two lemmas which do not depend on The Subgroup Theorem, but will be of use later.

Lemma 5.1 *These exists a recursively enumerable set $\{e_i \mid i \in \mathbb{N}\}$ of words in the variables x and y and their inverses such that, for any elements $g_0, g_1, g_2, \ldots$ in any group G, the equations $e_i = g_i$ $(i \in \mathbb{N})$ can be solved over G.*

Proof A result of Higman, Neumann, and Neumann (1949) states that any countable group can be embedded in a two-generator group. In fact, (see Higman *et al.*, 1949) there is an embedding of any given group $G = \langle g_0, g_1, \ldots \rangle$ in some group $H = \langle a, b \rangle$, which carries g_i to

$$a^{-1}b^{-1}ab^{-i}ab^{-1}a^{-1}b^{i}a^{-1}bab^{-i}aba^{-1}b^{i}$$

$(i \in \mathbb{N})$. Thus we put

$$e_i = x^{-1}y^{-1}xy^{-i}xy^{-1}x^{-1}y^{i}x^{-1}yxy^{-i}xyx^{-1}y^{i}.$$

Clearly, the set $\{e_i \mid i \in \mathbb{N}\}$ is recursive, and the equations $e_i = g_i$ can all be solved in $H \supseteq G$. □

Note If R is a set of defining relations for G, G can be embedded in $H = \langle a, b \mid r(e_0, e_1, \ldots) = 1, \quad r \in R \rangle$, by the map

$$g_i \mapsto e_i(a, b) = a^{-1}b^{-1}ab^{-i}ab^{-1}a^{-1}b^ia^{-1}bab^{-i}aba^{-1}b^i.$$

Lemma 5.2 *For elements* $a_1, \ldots, a_r, b_1, \ldots, b_r$, *of any group* G, *the following two statements are equivalent.*

(*i*) *There exists a homomorphism* $\theta: \langle a_1, \ldots, a_r \rangle \to \langle b_1, \ldots, b_r \rangle$ *such that* $a_i\theta = b_i$ $(1 \leqslant i \leqslant r)$.

(*ii*) *The equations* $a_ix^{-1}b_jx = x^{-1}b_jxa_i$ *and* $y^{-1}a_iy = a_ix^{-1}b_ix$ $(1 \leqslant i,j \leqslant r)$ *are soluble over* G.

Proof (i)$\Rightarrow$(ii). Let $C = \langle c_1, \ldots, c_r \rangle$ with

$$\text{typ}\,(b_1, \ldots, b_r) = \text{typ}\,(c_1, \ldots, c_r),$$

and form the direct product $G \times C$. Then, for all words of the form $w(a_1c_1, \ldots, a_rc_r)$ in $G \times C$, we have

$$w(a_1c_1, \ldots, a_rc_r) = w(a_1, \ldots, a_r)w(c_1, \ldots, c_r).$$

So $w(a_1c_1, \ldots, a_rc_r) = 1$ if and only if $w(\boldsymbol{a}) = w(\boldsymbol{c}) = 1$, which holds if and only if $1 = w(\boldsymbol{a})$ and $1 = \mathrm{w}(\boldsymbol{b}) = \mathrm{w}(\boldsymbol{b\theta})$, and hence if and only if $w(\boldsymbol{a}) = 1$. So

$$\text{typ}\,(a_1, \ldots, a_r) = \text{typ}\,(a_1c_1, \ldots, a_rc_r)$$

and we can construct the repeated HNN-extension

$$\langle G \times C, \bar{x}, \bar{y} \mid \bar{x}^{-1}b_i\bar{x} = c_i, \quad \bar{y}^{-1}a_i\bar{y} = a_ic_i, \quad 1 \leqslant i \leqslant r \rangle.$$

Then a_i and c_j commute for all i and j, so the equations in (ii) are soluble over G.

(ii)$\Rightarrow$(i). We require that $w(\boldsymbol{a}) = 1$ implies that $w(\boldsymbol{b}) = 1$, for any word w. But

$$\begin{aligned} \bar{y}^{-1}w(\boldsymbol{a})\bar{y} = w(\bar{y}^{-1}a_1y, \ldots, \bar{y}^{-1}a_ry) &= w(a_1\bar{x}^{-1}b_1\bar{x}, \ldots, a_r\bar{x}^{-1}b_r\bar{x}) \\ &= w(\boldsymbol{a})\bar{x}^{-1}w(\boldsymbol{b})\bar{x}, \end{aligned}$$

so the result holds. □

Let $G = \langle a_1, \ldots, a_r \rangle$ be a finitely generated group. By equations, inequalities, and positive and negative implications over G, we shall mean statements of the form $w = 1$, $w \neq 1$,

$$(u_1 = 1 \;\&\; \cdots \;\&\; u_k = 1) \Rightarrow (v = 1),$$

and

$$(u_1 = 1 \;\&\; \cdots \;\&\; u_k = 1) \Rightarrow \Omega,$$

respectively, where w, v, and u_i are words on the $a_1, \ldots, a_r, x_0, x_1, \ldots$ and their inverses, and Ω is some false statement. Strictly speaking, w, v, and u_i ought to be elements of the free product $G * F$, where $x_0, x_1, \dots$ freely generate F, but we want to talk about recursively enumerable subsets of equations etc., and G is not effectively enumerable if its word problem is unsolvable. So, if $\mathscr{S}$ is a set of equations, inequalities, and positive and negative implications over $G = \langle a_1, \dots, a_r \rangle$, by the statement '$\mathscr{S}$ is recursively enumerable' we shall mean that the set of all words w, v, and u_i appearing in elements of $\mathscr{S}$ is a recursively enumerable subset of the set of all words on the $a_1, \dots, a_r, x_0, x_1, \dots$ and their inverses.

Definitions A set $\mathscr{S}$ (as above) of equations, inequalities, and implications over $G = \langle a_1, \dots, a_r \rangle$ is *satisfiable over* G if there exists a group $L \supseteq G$, and a homomorphism

$$\varphi : G * F \to L,$$

such that:

(i) the restriction of φ to G is the identity,
(ii) if $w = 1$ belongs to $\mathscr{S}$ then $w\varphi = 1$,
(iii) if $w \neq 1$ belongs to $\mathscr{S}$ then $w\varphi \neq 1$,
(iv) if $(u_1 = 1 \;\&\; \cdots \;\&\; u_k = 1) \Rightarrow (v = 1)$ belongs to $\mathscr{S}$, and if

$$u_1\varphi = \cdots = u_k\varphi = 1$$

in L, then $v\varphi = 1$,
(v) if $(u_1 = 1 \;\&\; \cdots \;\&\; u_k = 1) \Rightarrow \Omega$ belongs to $\mathscr{S}$, and if

$$u_i\varphi = \cdots = u_{k-1}\varphi = 1,$$

then $u_k\varphi \neq 1$.

In this case we will call the elements $x_i\varphi$ *solutions* for $\mathscr{S}$ in L.

The set $\mathscr{S}$ is *satisfiable in* G if we can take $L = G$ in the above definition.

Suppose that $\mathscr{S}$ consists of equations and positive implications only. Let E be the free group freely generated by the variables in $\mathscr{S}$. (Note: we have therefore fixed a set of variables over which $\mathscr{S}$ is defined. Clearly, this set must include all the variables actually occurring in members of $\mathscr{S}$. In practice, this is usually precisely the set of variables that we will take to generate E. But sometimes, for example if $\mathscr{S}$ is empty, we may wish to suppose that $\mathscr{S}$ is defined over some larger set of variables, and take this set to generate E. This causes no problems, as long as we remember that $\mathscr{S}$ has been defined over some fixed and unalterable set of variables.) Consider the free product $G * E$, and let N be the smallest normal subgroup of $G * E$ such that:

(a) $w \in N$, for all equation $w = 1$ in $\mathscr{S}$,

(b) *if* $(u_1 = 1 \;\&\; \cdots \;\&\; u_k = 1) \Rightarrow (v = 1)$ belongs to $\mathscr{S}$, and if $u_1, \dots, u_k$ all belong to N, then $v \in N$.

We call $(G*E)/N$ *the group obtained by freely imposing the solutions of* $\mathcal{S}$ *on* G, and we denote it by $\langle G, \mathcal{S}\rangle$.

In general, $\langle G, \mathcal{S}\rangle$ need not contain G, but it will contain a natural homomorphic image of G, in the sense that there is a natural map

$$\alpha: G \to \langle G, \mathcal{S}\rangle,$$

and $G \subseteq \langle G, \mathcal{S}\rangle$ when α is a monomorphism, i.e. when $G \cap N = \{1\}$. Clearly then, $\mathcal{S}$ is satisfiable over G if and only if $G \cap N = \{1\}$ and in this case $\langle G, \mathcal{S}\rangle$ is the group obtained by adjoining the solutions of $\mathcal{S}$ to G. If the natural map α is an embedding, we will suppress it, and assume that $G \subseteq \langle G, \mathcal{S}\rangle$, i.e. $\alpha = 1$.

If $\mathcal{S}$ is satisfiable in $L \supseteq G$, and if $\varphi: G*E \to L$ is a homomorphism satisfying the conditions (i)–(v) above, then there is a natural homomorphism $\bar{\varphi}: \langle G, \mathcal{S}\rangle \to L$ such that the diagram

$$\begin{array}{ccc} G * E & \xrightarrow[\text{projection}]{\text{natural}} & (G * E)/N \\ & \searrow^{\varphi} & \downarrow{\scriptstyle \bar{\varphi}} \\ & & L \end{array}$$

commutes.

If $\mathcal{S}_1$ and $\mathcal{S}_2$ are two sets of equations and positive implications over G, and if

$$\alpha_i: G \to \langle G, \mathcal{S}_i\rangle \quad (1 \leq i \leq 2)$$

are the natural maps, then a homomorphism

$$\theta: \langle G, \mathcal{S}_1\rangle \to H \supseteq \langle G, \mathcal{S}_2\rangle$$

is said to be *G-preserving* if

$$(gN_1\theta =) \quad g\alpha_1\theta = g\alpha_2 \quad (= gN_2) \quad \text{for all } g \in G.$$

In particular, if there is a G-preserving homomorphism from $\langle G, \mathcal{S}_1\rangle$ to $\langle G, \mathcal{S}_2\rangle$, then $G \cap N_1 \neq \{1\}$ implies that $g\alpha_1 = 1$, for some $g \in G_1 \backslash \{1\}$, which implies that $g\alpha_2 = 1$. (Here, $\langle G, \mathcal{S}_i\rangle = (G*E_i)/N_i$.) So, if $\mathcal{S}_2$ is satisfiable over G, then $\mathcal{S}_1$ is satisfiable over G.

Further, if $\mathcal{S}_2$ is satisfiable in $L \supseteq G$, there is a homomorphism $\varphi: G*E_2 \to L$, satisfying (i)–(v) above; so we have the following diagram.

$$\begin{array}{ccccc} G*E_2 & \xrightarrow{\text{natural}} & \langle G, \mathcal{S}_2\rangle & \xrightarrow{\bar{\varphi}} & L \\ & & \uparrow{\scriptstyle \theta} & & \\ G*E_1 & \xrightarrow{\text{natural}} & \langle G, \mathcal{S}_1\rangle & & \end{array}$$

Letting φ' denote the homomorphism from $G * E_1$ to L given by this diagram, we see that, if $\mathscr{S}_1$ consists of equations only, then φ' satisfies (i)–(v), and hence that $\mathscr{S}_1$ is satisfiable in L.

The following theorem allows us to prove that, if $\mathscr{S}$ is any recursively enumerable set of equations, inequalities, and implications that is satisfiable over an existentially closed group M, then $\mathscr{S}$ is satisfiable in M.

Theorem 5.3 (see also Theorem 9.8) *Let G be a non-trivial finitely generated group, and let $\mathscr{S}$ be a recursively enumerable set of equations, inequalities, and positive and negative implications over G. Then there exists a finite set $\mathscr{S}^*$ of equations only, over G, such that*

(i) if $\mathscr{S}$ is satisfiable over G then $\mathscr{S}^*$ is soluble over G,

(ii) if $L \supseteq G$ and if $\mathscr{S}^*$ is soluble in L then $\mathscr{S}$ is satisfiable in L.

Proof We show that there is a sequence $\mathscr{S}_0$, $\mathscr{S}_1$, $\mathscr{S}_2$, $\mathscr{S}_3$, $\mathscr{S}_4$ of recursively enumerable sets such that

(a) $\mathscr{S} = \mathscr{S}_0$, and $\mathscr{S}^* = \mathscr{S}_4$ is a finite set of equations,

(b) $\mathscr{S}_1$ consists of equations and positive implications only,

(c) $\mathscr{S}_2$ consists of equations only,

(d) $\mathscr{S}_3$ consists of equations in two variables,

(e) if $\mathscr{S}_i$ is satisfiable over G, then so is $\mathscr{S}_{i+1}$ $(0 \leqslant i \leqslant 3)$,

(f) if $\mathscr{S}_{i+1}$ is satisfiable in $L \supseteq G$, then so is $\mathscr{S}_i$ $(0 \leqslant i \leqslant 3)$.

Clearly, the result follows from these.

To construct $\mathscr{S}_1$, choose $g \in G_1 \setminus \{1\}$, and replace the inequalities

$$w \neq 1,$$

and the negative implications

$$(u_1 = 1 \;\&\; \cdots \;\&\; u_k = 1) \Rightarrow \Omega,$$

in $\mathscr{S}$, by

$$(w = 1) \Rightarrow (g = 1),$$
$$(u_1 = 1 \;\&\; \cdots \;\&\; u_k = 1) \Rightarrow (g = 1),$$

respectively. Clearly, $\mathscr{S}_1$ is satisfiable in a group $H \supseteq G$ if and only if $\mathscr{S}$ is satisfiable in H, and $\mathscr{S}_1$ is recursively enumerable if $\mathscr{S}$ is recursively enumerable.

From Lemma 1.6, we see that, if the positive implication

$$(u_1 = 1 \;\&\; \cdots \;\&\; u_k = 1) \Rightarrow (v = 1) \tag{1}$$

holds in some group $H \supseteq G$, then the equation

$$x_1^{-1}u_1x_1y_1^{-1}u_1y_1 \cdots x_k^{-1}u_kx_ky_k^{-1}u_ky_k = v \tag{2}$$

is soluble over H and hence over G. Clearly, if the equation (2) is soluble in $L \supseteq G$, then the implication (1) holds in L. If $\mathscr{S}_2$ is the set of equations obtained from $\mathcal{S}_1$ by replacing positive implications by equations in this way, we see that $\mathcal{S}_2$ is recursively enumerable, and hence that $\mathcal{S}_2$ is the required such set.

For ease of notation, re-name so that $x_0, x_1, x_2, \ldots$ are the variables in $\mathcal{S}_2$, and choose, as in Lemma 5.1, a recursive sequence

$$e_0, e_1, e_2, \ldots$$

of words on the letters y and z and their inverses. Consider the set $\mathcal{S}_3$ of equations over G in the variables y and z that is obtained from $\mathcal{S}_2$ by substituting e_i for x_i $(i \in \mathbb{N})$. Clearly, if $\mathcal{S}_3$ can be solved in $L \supseteq G$, then $\mathcal{S}_2$ can be solved in L. If $\mathcal{S}_2$ can be solved over G, in H say, then the equations $\bar{x}_i = e_i(y, z)$ $(i \in \mathbb{N})$ (where $\bar{x}_i$ are solutions for $\mathcal{S}_2$ in H) can be solved over H, by Lemma 5.1. Hence $\mathcal{S}_3$ can be solved over H, and thus over G. If $\mathcal{S}_2$ is recursively enumerable, then so is $\mathcal{S}_3$, as required.

We now construct $\mathcal{S}^*$. Let

$$G = \langle a_1, \ldots, a_r \rangle,$$

and suppose that $\mathcal{S}_3$ is a set of equations in two variables x_1 and x_2, (instead of y and z). Take symbols $\hat{a}_i$ $(1 \leq i \leq r)$, $\hat{x}_1$, and $\hat{x}_2$, and let

$$H = \langle \hat{a}_1, \ldots, \hat{a}_r, \hat{x}_1, \hat{x}_2 \mid w(\hat{\boldsymbol{a}}, \hat{\boldsymbol{x}}) = 1;\ \text{whenever } w = 1 \text{ lies in } \mathcal{S}_3 \rangle,$$

so that $\langle G, \mathcal{S}_3 \rangle$ is a homomorphic image of H. Since $\mathcal{S}_3$ is recursively enumerable, we can use The Subgroup Theorem to embed H in some finitely presented group

$$K = \langle b_1, \ldots, b_t \mid r_1(\boldsymbol{b}) = \cdots = r_n(\boldsymbol{b}) = 1 \rangle.$$

Let $s_i(\boldsymbol{b})$ $(i \leq i \leq r)$, $s_{r+1}(\boldsymbol{b})$, and $s_{r+2}(\boldsymbol{b})$ be the images of $\hat{a}_i$ $(1 \leq i \leq r)$, $\hat{x}_1$, and $\hat{x}_2$, respectively, in K.

Take variables $x_1, x_2, y_1, \ldots, y_t, u, v$, let $s_i = s_i(\boldsymbol{y})$, and let $\mathcal{S}^*$ be the set

$$\left\{\begin{array}{c|c} r_p(\boldsymbol{y}) = 1 & 1 \leq p \leq n \\ s_q u^{-1} a_i u = u^{-1} a_i u s_q & 1 \leq q \leq r+2 \\ s_q u^{-1} x_j u = u^{-1} x_j u s_q & 1 \leq i \leq r \\ v^{-1} s_i v = s_i u^{-1} a_i u & 1 \leq j \leq 2 \\ v^{-1} s_{r+j} v = s_{r+j} u^{-1} a_i u & \end{array}\right\}$$

of equations over G. Let E_3 and E^* be the free groups freely generated by x_1 and x_2, and by $x_1, x_2, y_1, \ldots, y_t, u, v$, respectively. For $w \in G * E_3$, let $\bar{w}$ denote the image of w in $\langle G, \mathcal{S}_3 \rangle$ under the natural homomorphism

$$G * E_2 \to (G * E_3)/N_3.$$

So $\bar{w} = wN_3$ and, for $a \in G$, we have $\bar{a} = a\alpha_3$. Similarly, for $w \in G * E^*$, let $\bar{\bar{w}}$ be the image of w in $\langle G, \mathcal{S}^* \rangle$. From the discussion just above the statement of the theorem, we see that, if there is a G-preserving homomorphism

$$\theta:\langle G, \mathcal{S}_3\rangle \to \langle G, \mathcal{S}^*\rangle,$$

then $\mathcal{S}_3$ is satisfiable in any group $L \supseteq G$ in which $\mathcal{S}^*$ is satisfiable. Now, the rules

$$\bar{a}_i \mapsto \bar{\bar{a}}_i \quad (1 \leq i \leq r), \qquad \bar{x}_j \mapsto \bar{\bar{x}}_j \quad (1 \leq j \leq 2)$$

define such a (G-preserving) homomorphism if, for all words $w(\boldsymbol{a}, \boldsymbol{x}) \in \mathrm{Rel}\,(G)$, and for all equations $w(\boldsymbol{a}, \boldsymbol{x}) = 1$ in $\mathcal{S}_3$, the statement $(w(\bar{\bar{\boldsymbol{a}}}, \bar{\bar{\boldsymbol{x}}}) =)\ \bar{\bar{w}} = 1$ holds in $\langle G, \mathcal{S}^*\rangle$. Since $\mathrm{Rel}\,G \subseteq \mathrm{Rel}\,\langle G, \mathcal{S}^*\rangle$, we only need to consider the case where $w(\boldsymbol{a}, \boldsymbol{x}) = 1$ lies in $\mathcal{S}_3$.

The rules

$$b_l \mapsto \bar{\bar{y}}_l \quad (1 \leq l \leq t)$$

define a homomorphism $\varphi : K \mapsto \langle G, \mathcal{S}^*\rangle$, and the rules

$$\bar{\bar{y}}_l \mapsto b_l \quad (1 \leq l \leq t), \qquad \bar{\bar{x}}_j \mapsto 1 \quad (1 \leq j \leq 2),$$
$$\bar{\bar{a}}_1 \mapsto 1 \quad (1 \leq i \leq r), \qquad \bar{\bar{v}} \mapsto 1, \qquad \bar{\bar{u}} \mapsto 1,$$

define a homomorphism $\varphi' : \langle G, \mathcal{S}^*\rangle \to K$, and hence φ is a monomorphism. Now, $w(\boldsymbol{a}, \boldsymbol{x}) = 1$ lies in $\mathcal{S}_3$ if and only if $w(\boldsymbol{a}, \boldsymbol{x}) \in \mathrm{Rel}\,H$, which, since φ is an embedding, is if and only if

$$w(\bar{\bar{s}}_1, \ldots, \bar{\bar{s}}_{r+2}) = 1$$

holds in $\langle G, \mathcal{S}^*\rangle$. But, as in Lemma 5.2, the equations in $\mathcal{S}^*$ guarantee that there exists a homomorphism

$$\langle \bar{\bar{s}}_1, \ldots, \bar{\bar{s}}_{r+2}\rangle \to \langle \bar{\bar{a}}_1, \ldots, \bar{\bar{a}}_r, \bar{\bar{x}}_1, \bar{\bar{x}}_2\rangle$$

which maps $\bar{\bar{s}}_i$ to $\bar{\bar{a}}_i$ $(1 \leq i \leq r)$ and $\bar{\bar{s}}_{r+j}$ to $\bar{\bar{x}}_j$ $(1 \leq j \leq 2)$. So, if $w(\bar{\bar{s}}_1, \ldots, \bar{\bar{s}}_{r+2}) = 1$ holds in $\langle G, \mathcal{S}^*\rangle$, then so does $w(\bar{\bar{a}}, \bar{\bar{x}}) = 1$, and the above rules do define a G-preserving homomorphism.

Finally we want to show that if $\mathcal{S}_3$ is soluble over G, then so is $\mathcal{S}^*$.

Let L be the direct product

$$\langle G, \mathcal{S}_3\rangle \times K,$$

which contains G if $\mathcal{S}_3$ is soluble over G. The relations

$$\{r_p(\boldsymbol{b}) = 1 \mid 1 \leq p \leq n\}$$

hold in L, and there is a homomorphism

$$\langle s_1(\boldsymbol{b}), \ldots, s_{r+2}(\boldsymbol{b})\rangle \to \langle G, \mathcal{S}_3\rangle$$

defined by:

$$s_i(\boldsymbol{b}) \to \bar{a}_i \quad (1 \leq i \leq r), \qquad s_{r+j}(\boldsymbol{b}) \to \bar{x}_j \quad (1 \leq j \leq 2).$$

(Recall that $H \cong \langle s_1(\boldsymbol{b}), \ldots, s_{r+2}(\boldsymbol{b})\rangle$, and so $w(s_1, \ldots, s_{r+2}) = 1$ if and only if $w(\boldsymbol{a}, \boldsymbol{x}) \in \mathcal{S}_3$). So, by Lemma 5.2, the relations $\mathcal{S}^*$ are soluble over $L \supseteq G$, as required, and we have proved the theorem. □

Note We used the fact the G was non-trivial in the above theorem only to prove the existence of the recursively enumerable set $\mathcal{S}_1$. Thus, if $\mathcal{S}$ is already a recursively enumerable set of equations and positive implications only, we do not need to assume that $G \neq 1$. Thus the proof of Theorem 5.3 gives us the following result.

Theorem 5.4 *If $\mathcal{S}$ is a recursively enumerable set of equations and positive implications defined over any finitely generated group G, then there exists a finite set $\mathcal{S}^*$, of equations over G, such that*
 (i) if $\mathcal{S}$ is satisfiable over G, then $\mathcal{S}^*$ is satisfiable over G,
 (ii) if $\mathcal{S}^*$ is satisfiable in $L \supseteq G$, then $\mathcal{S}$ is satisfiable in L.

Putting $L = M$ in Theorem 5.3, we obtain the following direct corollary.

Corollary 5.5 *If $\mathcal{S}$ is a recursively enumerable set of equations, inequalities, and positive and negative implications, defined over any finitely generated subgroup of an existentially closed group M, and satisfiable over M, then $\mathcal{S}$ is satisfiable in M.*

(Note: We do not need the assumption that $\mathcal{S}$ is non-trivial in the corollary because, if $\mathcal{S}$ is defined over 1, then it is clearly defined over $\langle g \rangle$, for any $g \in M \backslash \{1\}$.)

The next theorem is also a corollary of Theorem 5.3, and it will be needed in Chapter 6.

Theorem 5.6 *If $\mathcal{S}$ is a recursively enumerable set of equations and positive implications defined over a finitely generated group G, and if $\mathcal{S}^*$ is a corresponding finite set of equations constructed as in the proof of Theorem 5.3, then there is a G-preserving embedding*

$$\theta : \langle G, \mathcal{S}\rangle \to \langle G, \mathcal{S}^*\rangle.$$

Proof We aim to show that for $\mathcal{S}_1$, $\mathcal{S}_2$, $\mathcal{S}_3$, and $\mathcal{S}_4$ as in the proof of Theorem 5.3, there exist G-preserving homomorphisms

$$\theta_2 : \langle G, \mathcal{S}_i\rangle \to \langle G, \mathcal{S}_{i+1}\rangle$$

which are embeddings. Then, since we can take $\mathcal{S} = \mathcal{S}_1$ and $\mathcal{S}^* = \mathcal{S}_4$, the map $\theta = \theta_1\theta_2\theta_3$ is the required G-preserving embedding.

First we make a preliminary remark.

Suppose that $\mathcal{S}$ is defined over the variables

$$\{x_i, y_j \mid i \in I, \quad j \in J\},$$

and that $\mathcal{S}'$ is another set of equations and positive implications over G, defined over the variables

$$\{x_i, z_k \mid i \in I, \quad k \in K\},$$

where $y_j \neq z_k$, for any $j \in J$ and $k \in K$. We can associate with $\mathcal{S}'$ a set $\mathcal{S}'(N)$ of equations and implications defined over the group $\langle G, \mathcal{S}\rangle = (G * E)/N$, by replacing g by gN, for all $g \in G$, and x_i by x_iN, for all $i \in I$, in $\mathcal{S}'$. (Here E is freely generated by $\{x_i, y_j \mid i \in I, \quad j \in J\}$.) It is straighforward to check from the definition of $\langle G, \mathcal{S}'\rangle$ that *if* $\mathcal{S}'(N)$ *is satisfiable in* $H \supseteq \langle G, \mathcal{S}\rangle$, *and if* $\bar{z}_k$ $(k \in K)$ *are solutions for* $\mathcal{S}'(N)$ *in* H, *then the rules*

$$gN' \mapsto gN \quad (g \in G), \qquad x_iN' \mapsto x_iN \quad (i \in I), \qquad z_kN' \mapsto \bar{z}_k \quad (k \in K)$$

define a G-preserving homomorphism from $\langle G, \mathcal{S}'\rangle$ *to* H. (Here E' is freely generated by $\{x_i, z_k \mid i \in I, \quad k \in K\}$ and $\langle G, \mathcal{S}'\rangle = (G * E')/N'$.)

Now, let $\mathcal{S} = \mathcal{S}_1$ be our recursively enumerable set of equations and positive implications, and let $\mathcal{S}_2$, $\mathcal{S}_3$, and $\mathcal{S}_4 = \mathcal{S}^*$, be sets constructed as in Theorem 5.3. Let E_i denote the free group freely generated by the variables in $\mathcal{S}_i$, then $\langle G, \mathcal{S}_i\rangle = (G * E_i)/N_i$ $(1 \leqslant i \leqslant 4)$.

All the variables in $\mathcal{S}_1$ are also in $\mathcal{S}_2$, so $\mathcal{S}_1(N_2)$ is a set of statements about elements of $\langle G, \mathcal{S}_2\rangle$. Since every equation in $\mathcal{S}_1$ is also in $\mathcal{S}_2$, all the equalities in $\mathcal{S}_1(N_2)$ hold in $\langle G, \mathcal{S}_2\rangle$. For each implication

$$(u_1 = 1 \,\&\, \cdots \,\&\, u_k = 1) \Rightarrow (v = 1)$$

in $\mathcal{S}_1$, there is a corresponding equation

$$x_1^{-1}u_1x_1y_1^{-1}u_1y_1 \cdots x_k^{-1}u_kx_ky_k^{-1}u_ky_k = v$$

in $\mathcal{S}_2$. So the corresponding implication in $\mathcal{S}_1(N_2)$ holds in $\langle G, \mathcal{S}_2\rangle$. Thus $\mathcal{S}_1(N_2)$ is soluble in $\langle G, \mathcal{S}_2\rangle$, and hence, from the above discussion, we see that there exists a homomorphism

$$\theta_1 : \langle G, \mathcal{S}_1\rangle \to \langle G, \mathcal{S}_2\rangle$$

given by:

$$wN_1 \mapsto wN_2 \quad \text{for all } w \in G * E_1.$$

Conversely, those equations in $\mathcal{S}_2(N_1)$ that correspond to equations in $\mathcal{S}_1$ hold in $\langle G, \mathcal{S}_1\rangle$, and those equations in $\mathcal{S}_2(N_1)$ that correspond to implications in $\mathcal{S}_1$ are satisfiable over $\langle G, \mathcal{S}_1\rangle$, in H_1 (say), by Lemma 1.6 (because the corresponding implications hold in $\langle G, \mathcal{S}_1\rangle$). So, again from above, we see that there is a homomorphism

$$\varphi_1 : \langle G, \mathcal{S}_2 \rangle \to H_1$$

given by

$$wN_2 \mapsto wN_1 \quad (w \in G * E_1) \qquad zN_2 \mapsto \bar{z} \quad (z \in E_2 \backslash E_1).$$

Thus θ_1 is a (G-preserving) embedding.

Since the equation $w(x_0, x_1, \ldots) = 1$ belongs to $\mathcal{S}_2$ if and only if the equation $w(e_0, e_1, \ldots) = 1$ belongs to $\mathcal{S}_3$, the map

$$\theta_2 : \langle G, \mathcal{S}_2 \rangle \to \langle G, \mathcal{S}_3 \rangle$$

given by

$$gN_2 \mapsto gN_3 \quad (g \in G), \qquad x_i N_2 \mapsto e_i N_3 \quad (i \in \mathbb{N})$$

is clearly a G-preserving embedding.

We have seen, in the proof of Theorem 5.3, that there is a G-preserving homomorphism

$$\theta_3 : \langle G, \mathcal{S}_3 \rangle \to \langle G, \mathcal{S}_4 \rangle$$

given by

$$gN_3 \to gN_4 \quad (g \in G), \qquad x_i N_3 \to x_i N_4 \quad (1 \leqslant i \leqslant 2).$$

Also, as in the proof of Theorem 5.3, we see that the equations $\mathcal{S}_4(N_3)$ over $\langle G, \mathcal{S}_3 \rangle$, are soluble over $\langle G, \mathcal{S}_3 \rangle \times K$. So, again by the preliminary remark above, there is a homomorphism

$$\varphi_3 : \langle G, \mathcal{S}_4 \rangle \to \langle G, \mathcal{S}_3 \rangle$$

to which θ_3 is a partial inverse, i.e. φ_3 maps gN_4 to gN_3, for $g \in G$, and $x_i N_4$ to $x_i N_3$ $(1 \leqslant i \leqslant 2)$. So θ_3 is an embedding, and we have the result. □

We can now prove one half of an important theorem, mentioned in Chapter 4.

Although it is not necessarily true that if $G \in \operatorname{Sk} M$ and $\operatorname{Rel} H \leqslant_{\mathrm{T}} \operatorname{Rel} G$ then $H \in \operatorname{Sk} M$ we have the following result.

Theorem 5.7 *Let G and H be finitely generated groups with $\operatorname{Rel} H \leqslant^* \operatorname{Rel} G$. Then, for any existentially closed group M, if $G \in \operatorname{Sk} M$ we have $H \in \operatorname{Sk} M$.*

Proof Let $H = \langle b_1, \ldots, b_s \rangle$ and let $G = \langle a_1, \ldots, a_r \rangle \subseteq M$. Let $\mathrm{W}(\boldsymbol{x})$ and $\mathrm{W}(\boldsymbol{y})$ be the sets of words on the variables $x_1, \ldots, x_r$ and $y_1, \ldots, y_s$ (and their inverses) respectively. We shall think of $\operatorname{Rel}(a_1, \ldots, a_r)$ and $\operatorname{Rel}(b_1, \ldots, b_s)$ as lying in $\mathrm{W}(\boldsymbol{x})$ and $\mathrm{W}(\boldsymbol{y})$, respectively. Now,

$$\operatorname{Rel}(b_1, \ldots, b_s) \leqslant^* \operatorname{Rel}(a_1, \ldots, a_r);$$

so there exist recursively enumerable subsets

$$U \subseteq \mathrm{W}(\boldsymbol{y}) \times \mathscr{P}_{\mathrm{f}}(\mathrm{W}(\boldsymbol{x})), \qquad V \subseteq \mathrm{W}(\boldsymbol{y}) \times \mathscr{P}_{\mathrm{f}}(\mathrm{W}(\boldsymbol{x})) \times (\mathrm{W}(\boldsymbol{x}) \cup \{\infty\}),$$

such that

$$\mathrm{Rel}\,(b_1, \ldots, b_s) \leqslant_{\mathrm{e}} \mathrm{Rel}\,(a_1, \ldots, a_r) \quad \text{via } U$$

and

$$w(\boldsymbol{y}) \in \mathrm{Nonrel}\,(b_1, \ldots, b_s) = \mathrm{W}(\boldsymbol{y}) \setminus \mathrm{Rel}\,(b_1, \ldots, b_s),$$

if and only if there exist

$$A \subseteq \mathrm{Rel}\,(a_1, \ldots, a_r) \quad \text{and} \quad V \in \mathrm{Nonrel}\,(a_1, \ldots, a_r) \cup \{\infty\}$$

such that $(w, A, v) \in V$. From U and V, we form a set $\mathscr{S}$ of implications as follows. For each $(w, A) \in U$, take the implication

$$(u_1(\boldsymbol{a}) = 1 \,\&\, \cdots \,\&\, u_k(\boldsymbol{a}) = 1) \Rightarrow (w(\boldsymbol{y}) = 1);$$

for each $(w, A, v) \in V$, take the implication

$$(u_1(\boldsymbol{a}) = 1 \,\&\, \cdots \,\&\, u_k(\boldsymbol{a}) = 1 \,\&\, w(\boldsymbol{y}) = 1) \Rightarrow (v(\boldsymbol{a}) = 1);$$

and for each $(w, A, \infty) \in V$, take the implication

$$(u_1(\boldsymbol{a}) = 1 \,\&\, \cdots \,\&\, u_k(\boldsymbol{a}) = 1 \,\&\, w(\boldsymbol{y}) = 1) \Rightarrow \Omega$$

(where $A = \{u_1(\boldsymbol{x}), \ldots, u_k(\boldsymbol{x})\}$).

Then $\mathscr{S}$ is a recursively enumerable set, which is satisfiable in $M \times H$, and hence, by Corollary 5.5, in M. If $\bar{b}_1, \ldots, \bar{b}_s \in M$ are solutions for $\mathscr{S}$, then, for all $w \in \mathrm{Rel}\,(b_1, \ldots, b_s)$, there exists

$$\{u_1, \ldots, u_k\} \subseteq \mathrm{Rel}\,(a_1, \ldots, a_r)$$

such that the implication

$$(u_1(\boldsymbol{a}) = 1 \,\&\, \cdots \,\&\, u_k(\boldsymbol{a}) = 1) \Rightarrow (w(\boldsymbol{y}) = 1)$$

belongs to $\mathscr{S}$. So $w(\bar{b}_1, \ldots, \bar{b}_s) = 1$ and thus

$$\mathrm{Rel}\,(b_1, \ldots, b_s) \subseteq \mathrm{Rel}\,(\bar{b}_1, \ldots, \bar{b}_s).$$

Similarly,

$$\mathrm{Nonrel}\,(b_1, \ldots, b_s) \subseteq \mathrm{Nonrel}\,(\bar{b}_1, \ldots, \bar{b}_s).$$

So, since

$$\mathrm{Rel}\,(b_1, \ldots, b_s) \cup \mathrm{Nonrel}\,(b_1, \ldots, b_s) = \mathrm{W}(\boldsymbol{y}),$$

we have

$$\mathrm{typ}\,(b_1, \ldots, b_s) = \mathrm{typ}\,(\bar{b}_1, \ldots, \bar{b}_s),$$

and hence $\langle \bar{b}_1, \ldots, \bar{b}_s \rangle \cong H$, giving the result. □

The next theorem gives the sufficient half of a necessary and sufficient condition for a finitely generated group to be embeddable in every existentially closed group.

Theorem 5.8 *A finitely generated group with solvable word problem is embeddable in every existentially closed group M.*

Proof Suppose that $G = \langle a_1, \dots, a_r \rangle$ has solvable word problem. Take variables $x_1, \dots, x_r$ and let

$$\mathcal{S} = \{w(\boldsymbol{x}) = 1 \mid w(\boldsymbol{a}) = 1\} \cup \{u(\boldsymbol{x}) \neq 1 \mid u(\boldsymbol{a}) \neq 1\}.$$

Since G has solvable word problem, $\mathcal{S}$ is recursively enumerable. By Corollary 5.5, since $\mathcal{S}$ is soluble in $M * G$, we have that $\mathcal{S}$ is soluble in M, and the solutions clearly generate a subgroup of M isomorphic to G. So $G \in \operatorname{Sk} M$. □

We might ask if there exists a 'minimum' existentially closed group, i.e. an existentially closed group whose skeleton consists precisely of the finitely generated groups with solvable word problem. To answer this question, we first need the following lemma.

Lemma 5.9 *Let A and B be disjoint recursively enumerable subsets of* $\mathbb{N}$, *and let M be an existentially closed group. Then* $\mathrm{G}_X \in \operatorname{Sk} M$, *for some X such that* $A \subseteq X$ *and* $B \cap X = \varnothing$.

Proof Take variables x and y, let $y_i = x^{-i} y x^i$, and let $z_i = [y, y_{i+1}]$ ($i \in \mathbb{N}$). We take equations $z_j = 1$, for all $j \in A$. Let w be a word in x, x^{-1}, y, y^{-1}, and suppose that the normal form of w is $x^\alpha y_{i_1}^{\beta_1} \cdots y_{i_k}^{\beta_k} z_{j_1} \cdots z_{j_l}$. We take the equation $w = 1$, if $\alpha = k = l = 0$, and the inequality $w \neq 1$, if $\alpha \neq 0$ or $k \neq 0$. We also take the inequalities $z_i \neq 1$, for all $i \in B$, and all the implications

$$(z_{j_1} z_{j_2} \cdots z_{j_l} = 1) \Rightarrow (z_{j_1} = 1)$$

for which $j_1 < j_2 < \cdots < j_l$ and $l \geq 2$. The set of all equations, inequalities, and implications thus obtained is recursively enumerable, and is satisfiable in $M \times \mathrm{G}_A$. Hence we can find solutions, a and b say, in M, and we see that $\langle a, b \rangle \cong \mathrm{G}_X$, for some X such that $A \subseteq X$ and $B \cap X = \varnothing$. □

Theorem 5.10 *Every existentially closed group M contains a finitely generated group with unsolvable word problem.*

Proof Take sets X and Y, as in Lemma 4.20. By Lemma 5.9, $\mathrm{G}_U \in \operatorname{Sk} M$, for some U with $X \subseteq U$ and $Y \cap U = \varnothing$. If U is recursively enumerable then, by Lemma 4.20, U is strongly creative and hence not recursive. So either $\mathbb{N} \backslash U$ or U is not recursively enumerable, and hence G_U does not have solvable word problem. □

The above theorem will be used in Chapter 7 as part of a proof that there exist $2^{\aleph_0}$ distinct countable existentially closed groups, whose skeletons intersect precisely in the finitely generated groups with solvable word problem. This will provide the matching necessary half of the condition in Theorem 5.8.

Although there is no minimum existentially closed group, the next theorem shows that $\mathrm{M}_\varnothing$ is minimal in the sense that it has a minimal skeleton.

Theorem 5.11 *If M is any existentially closed group such that* $\mathrm{Sk}\, M \subseteq \mathrm{Sk}\, \mathrm{M}_\varnothing$, then $\mathrm{Sk}\, M = \mathrm{Sk}\, \mathrm{M}_\varnothing$. *In particular, if M is also countable, then* $M \cong \mathrm{M}_\varnothing$.

Proof Again, take sets X and Y, as in Lemma 4.20, and, by Lemma 5.9, take U such that $X \subseteq U$, $U \cap Y = \varnothing$, and $\mathrm{G}_U \in \mathrm{Sk}\, M$. Then $\mathrm{G}_U \in \mathrm{Sk}\, \mathrm{M}_\varnothing$, by hypothesis, and so $\mathrm{Rel}\, \mathrm{G}_U$ is recursively enumerable. Hence, by Theorem 4.23, U is recursively enumerable and, by Lemma 4.20 and Lemma 4.22, U is 1-complete. For any $H \in \mathrm{Sk}\, \mathrm{M}_\varnothing$, we have that $\mathrm{Rel}\, H$ is recursively enumerable, and so $\mathrm{Rel}\, H \leqslant_1 U$. Then $\mathrm{Rel}\, H \leqslant^* U \leqslant^* \mathrm{Rel}\, \mathrm{G}_U$, and so, by Theorem 5.7, $H \in \mathrm{Sk}\, M$, giving the result. □

Remark To prove Theorem 5.11 we have used The Subgroup Theorem heavily. But conversely, we can deduce The Subgroup Theorem from Theorem 5.11 as follows. By Theorem 3.10 there exists an existentially closed group M such that $\mathrm{Sk}\, M$ is the set of all finitely generated subgroups of all finitely presented groups. Then each $G \in \mathrm{Sk}\, M$ has a recursively enumerable relation set, so $\mathrm{Sk}\, M \subseteq \mathrm{Sk}\, \mathrm{M}_\varnothing$. Thus, by Theorem 5.11, $\mathrm{Sk}\, M = \mathrm{Sk}\, \mathrm{M}_\varnothing$. If G is a finitely generated group with a recursively enumerable relation set, then $G \in \mathrm{Sk}\, \mathrm{M}_\varnothing$ and so $G \in \mathrm{Sk}\, M$. Thus, G belongs to some finitely presented group.

The next theorem was proved by Ziegler (1980) as a corollary of different results than those that we use to prove it here.

Theorem 5.12 *If M is an existentially closed group and if $G, H \in \mathrm{Sk}\, M$, then the direct product $G \times H$ belongs to* $\mathrm{Sk}\, M$.

Proof Let $G = \langle a_1, \dots, a_r \rangle$, let $H = \langle b_1, \dots, b_s \rangle$, and suppose that these are actually subgroups of M. Let $\mathscr{S}$ be the set of all equations of the form $a_i x^{-1} b_j x = x^{-1} b_j x a_i$ $(1 \leqslant i \leqslant r,\ \ 1 \leqslant j \leqslant s)$ and all implications of the form $(u(\boldsymbol{a}) = x^{-1} v(\boldsymbol{b}) x) \Rightarrow (u(\boldsymbol{a}) = 1)$, for all $u \in \mathrm{W}(x_1, \dots, x_r)$ and $v \in \mathrm{W}(y_1, \dots, y_s)$. Then $\mathscr{S}$ is recursively enumerable, and is satisfiable in the HNN-extension

$$\langle M \times H^*, x \mid x^{-1} b_i x = b_i^* \ (1 \leqslant i \leqslant s), \quad H^* \cong H \rangle.$$

By Corollary 5.5, we may take $x \in M$, and then $\langle G, x^{-1}Hx \rangle \cong G \times H$, as required. □

We remark here that, in general, the proof of this theorem requires, via Corollary 5.5, The Subgroup Theorem. However, if we weaken the result to consider only the case where G (or H) has trivial centre, or where $Z(G)$ and $Z(H)$ are finite, then the result follows directly from the definition of M, since we can take $\mathcal{S}$ to be a finite set of equations.

Before stating the next lemma, we define exactly what we mean by a wreath product. If H and G are abstract groups, we consider a set of groups $\{H^g \mid g \in G\}$, which are all disjoint and are all isomorphic to H. Form the direct product K of all the H^g. Clearly, the rule $h^g g_1 = h^{gg_1}$, $\forall g, g_1 \in G$, defines an action of G on K. Using this action, form a semidirect product of K with G. This semidirect product is called the (standard, restricted) *wreath product* of H with G, and it is written $H \operatorname{wr} G$. If H and G are subgroups of the same group L, then $\langle G, H \rangle$ is isomorphic to $H \operatorname{wr} G$ if and only if the conjugates $g^{-1}Hg$ of H by elements of G generate their direct product K, and $K \cap G = \{1\}$. In general, groups $B_1, B_2, \ldots$ in L generate their direct product if and only if $[b_i, b_j] = 1$, for all $i \neq j$, $b_i \in B_i$, and $B_i \cap \langle B_j \mid j \neq i \rangle = \{1\}$. Thus, the conjugates $g^{-1}Hg$ $(g \in G)$ generate their direct product K if and only if, for all $g \in G \backslash \{1\}$ and $h_1, h_2 \in H$, we have $h_1 g^{-1} h_2 g = g^{-1} h_2 g h_1$ and $H \cap \langle g^{-1}Hg \mid g \in G \backslash \{1\} \rangle = \{1\}$. This is so if and only if the equations

$$h_1 g^{-1} h_2 g = g^{-1} h_2 g h_1 \tag{1}$$

and the implications

$$(h_0^{-1} g_1^{-1} h_1 g_1 g_2^{-1} h_2 g_2 \cdots g_k^{-1} h_k g_k = 1) \Rightarrow (h_0 = 1) \tag{2}$$

hold in L, for $\{h_0, h_1, \ldots, h_k\} \subseteq H$, for $g_1, g_2, \ldots, g_k$ distinct elements of G, and for all k. Since every element of K can be written uniquely (up to permutation) in the form $g_1^{-1} h_1 g_1 g_2^{-1} h_2 g_2 \cdots g_k^{-1} h_k g_k$, where $g_1, \ldots, g_k$ are distinct elements of G and $\{h_1, \ldots, h_k\} \subseteq H$, then $K \cap G = \{1\}$ if and only if $g_1^{-1} h_1 g_1 \cdots g_k^{-1} h_k g_k \in G$ implies that $h_1 = h_2 = \cdots = h_k = 1$. If $g_1^{-1} h_1 g_1 \cdots g_k^{-1} h_k g_k \in G$ implies that $h_1 = 1$, for all distinct $g_1, \ldots, g_k$ in G, then $g_2^{-1} h_2 g_2 \cdots g_k^{-1} h_k g_k \in G$ and hence $h_2 = 1$. So $K \cap G = 1$ if and only if the implications

$$(g_1^{-1} h_1 g_1 g_2^{-1} h_2 g_2 \cdots g_k^{-1} h_k g_k = g_0) \Rightarrow (h_1 = 1) \tag{3}$$

hold in L, for each $h_i \in H$ and $g_0 \in G$, for all distinct elements $g_1, \ldots, g_k$ in G, and for all k. Note that the implications in (2) are a subset of the implications in (3).

Theorem 5.13 *If M is an existentially closed group, if G is a finitely generated group with solvable word problem, and if $H \in \operatorname{Sk} M$, then $H \operatorname{wr} G \in \operatorname{Sk} M$.*

Proof We may assume, by Theorem 5.8, that H and G lie in M. Let L be any group containing M. For each $\bar{x} \in L$ we see, from the above discussion, that H and $\bar{x}^{-1}G\bar{x}$ generate their wreath product $H \operatorname{wr} (\bar{x}^{-1}G\bar{x})$ in L, if and only if $\bar{x}$ satisfies the set $\mathscr{S}$, which consists of all equations and implications of the form

$$[h_i, x^{-1}g^{-1}xh_2x^{-1}gx] = 1,$$
$$(x^{-1}g_1^{-1}xh_1x^{-1}g_1x \cdots x^{-1}g_k^{-1}xh_kx^{-1}g_kx = x^{-1}g_0x) \Rightarrow (h_1 = 1),$$

where $g \in G\backslash\{1\}$, $g_0 \in G$, $\{h_1, \ldots, h_k\} \subseteq H$, $g_1, \ldots, g_k$ are distinct elements of G, and $k \in \mathbb{N}$.
The elements of H are effectively enumerable and, since G has solvable word problem, the set of all k-tuples ($k \in \mathbb{N}$) of distinct elements of G is recursive, so $\mathscr{S}$ is recursively enumerable. Let $H \operatorname{wr} G$ be the abstract wreath product and, for convenience of notation, denote by $\bar{g}$ the element of $H \operatorname{wr} G$ corresponding to $g \in G$. We can form the free product $M *_H (H \operatorname{wr} G)$, of M and $H \operatorname{wr} G$, amalgamating H. Consider the HNN-extension

$$L = \langle M *_H (H \operatorname{wr} G), \bar{x} \mid \bar{x}^{-1}g\bar{x} = \bar{g}, \quad g \in G \rangle.$$

Then $L \supseteq M$ and $\bar{x}$ satisfies $\mathscr{S}$; so, by Corollary 5.5, we can choose $t \in M$ to satisfy $\mathscr{S}$. Thus $H \operatorname{wr} (t^{-1}Gt) \subseteq M$, and clearly $H \operatorname{wr} (t^{-1}Gt)$ is isomorphic to the abstract wreath product, $H \operatorname{wr} G$. So $H \operatorname{wr} G \in \operatorname{Sk} M$, as required. □

Theorem 5.14 *If G is a finitely generated group with solvable word problem and if M is a countable existentially closed group, then M* wr *G can be embedded in M.*

Proof Any finitely generated subgroup of $M \operatorname{wr} G$ is a subgroup of some $H \operatorname{wr} G$, where $H \in \operatorname{Sk} M$. So, by Lemma 5.13, $\operatorname{Sk}(M \operatorname{wr} G) \subseteq \operatorname{Sk} M$. Then, by Lemma 3.4(i), $M \operatorname{wr} G$ can be embedded in M. □

Theorem 5.15 *If G is a finitely generated subgroup of the existentially closed group M, then there exists a finitely generated perfect group H such that $G \subseteq H \subseteq M$ and* $Z(H) = 1$.

Proof By Theorem 1.12, we can find a finitely generated perfect group $K \subseteq M$ such that $G \subseteq K$. Let

$$L = \langle a, b \mid a^2 = b^3 = (ab)^7 = 1 \rangle,$$

so that $L = L'$, the derived subgroup of L, and L has solvable word problem. By Theorem 5.8, we can assume that $L \subseteq M$. By Theorem 5.13, $H = K \operatorname{wr} L$ can be taken to be in M. From the properties of L and K, it is clear that H is finitely generated and perfect. If $s \in Z(K \operatorname{wr} L)$, where

$$s = g_n^{-1}k_ng_n \cdots g_1^{-1}k_1g_1g_0,$$

with $g_i \in t^{-1}Lt$ and $t \in M$, then

$$g_j^{-1}kg_j = s^{-1}g_j^{-1}kg_js = g_0^{-1}g_j^{-1}k_j^{-1}kk_jg_jg_0 \qquad (\forall k \in K),$$

and so, since $Kg_j \cap Kg_jg_0 \neq 1$, we have $g_0 = 1$. Since L is infinite we can choose $l \in t^{-1}Lt$ such that $l \neq g_i^{-1}g_j$, for any $i, j \in \{1, \ldots, n\}$. Then the fact that $sls^{-1} = l$ implies that

$$g_n^{-1}k_ng_n \cdots g_1^{-1}k_1g_1lg_1^{-1}k_1^{-1}g_1l^{-1}lg_2^{-1}k_2^{-1}g_2l^{-1} \cdots lg_n^{-1}k_ng_nl^{-1} = 1$$

and, since $g_1, \ldots, g_n, g_1l^{-1}, \ldots, g_nl^{-1}$ are all distinct, that $k_i = 1$ for $1 \leqslant i \leqslant n$. So $Z(H) = 1$, and H is the required group. □

6
The Relative-Subgroup Theorem

In this chapter we consider groups that can be obtained from a given group G by adjoining to G the solutions of some finite set of equations which are defined over G and are soluble over G. As a consequence of such a consideration, we are able to show, at the end of the chapter, that no existentially closed group is embeddable in a finitely presented group.

Definition 6.1 A group K is said to be *finitely presented over* G if K is isomorphic to $(G * E)/N$, where E is some finitely generated free group, N is the normal closure, in $G * E$, of a finite subset of $G * E$, and $N \cap G = 1$. (In other words, if K can be obtained by adjoining to G the solutions of a finite set of soluble equations over G.)

We now prove the main theorem of this section.

Theorem 6.2 (The Relative-Subgroup Theorem) *If G and H are finitely generated groups, then H is embeddable in a group that is finitely presented over G if and only if*

$$\operatorname{Rel} H \leq_e \operatorname{Rel} G.$$

Proof ($\Leftarrow$) The aim is to construct a set $\mathcal{S}$ such that $H \subseteq \langle G, \mathcal{S} \rangle$, and then to use Theorem 5.4 and Theorem 5.6 to embed H in $\langle G, \mathcal{S}^* \rangle$, where $\mathcal{S}^*$ is a finite set of equations over G.

Let $G = \langle a_1, \dots, a_r \rangle$, let $H = \langle b_1, \dots, b_s \rangle$, and let $\{x_1, \dots, x_r\}$ and $\{y_1, \dots, y_s\}$ be disjoint sets of variables. Then, if

$$\operatorname{Rel}(b_1, \dots, b_s) \leq_e \operatorname{Rel}(a_1, \dots, a_r),$$

by definition there exists a recursively enumerable set U, contained in

$$\mathrm{W}(y_1, \dots, y_s) \times \mathcal{P}_f(\mathrm{W}(x_1, \dots, x_r)),$$

such that

$$\operatorname{Rel}(b_1, \dots, b_s) \leq_e \operatorname{Rel}(a_1, \dots, a_r) \text{ via } U.$$

We form a set $\mathcal{S}$ of positive implications by taking

$$(u_1(\boldsymbol{a}) = 1 \;\&\; \cdots \;\&\; u_k(\boldsymbol{a}) = 1) \Rightarrow (v(\boldsymbol{y}) = 1),$$

whenever $(v(\boldsymbol{y}), \{u_1(\boldsymbol{x}), \ldots, u_k(\boldsymbol{x})\})$ belongs to U. Then $\mathscr{S}$ is recursively enumerable, and is satisfiable in the direct product $G \times H$. Let $\mathscr{S}^*$ be a corresponding finite set of equations, as in Theorem 5.4. Since $\mathscr{S}$ is soluble over G, the set $\mathscr{S}^*$ is soluble over G, and thus the group $\langle G, \mathscr{S}^* \rangle$ is finitely presented over G. By Theorem 5.6, $\langle G, \mathscr{S} \rangle$ is embedded in $\langle G, \mathscr{S}^* \rangle$.

We identify G with its natural images in $\langle G, \mathscr{S} \rangle$ and $\langle G, \mathscr{S}^* \rangle$.

By construction, there are solutions $\bar{y}_1, \ldots, \bar{y}_s$ for $\mathscr{S}$ in $\langle G, \mathscr{S} \rangle$. (If $\langle G, \mathscr{S} \rangle = (G * E)/N$, we can take $\bar{y}_i = y_i N$.) If $v(\boldsymbol{y}) \in \operatorname{Rel}(b_1, \ldots, b_s)$, then, by definition of $\mathscr{S}$, there is an implication

$$(u_1(\boldsymbol{a}) = 1 \ \& \cdots \& \ u_k(\boldsymbol{a}) = 1) \Rightarrow (v(\boldsymbol{y}) = 1)$$

in $\mathscr{S}$, where $u_i(\boldsymbol{x}) \in \operatorname{Rel}(a_1, \ldots, a_r)$. Since the equations $u_i(\boldsymbol{a}) = 1$ hold in $\langle G, \mathscr{S} \rangle$, and $\bar{y}_1, \ldots, \bar{y}_s$ are solutions for $\mathscr{S}$, we must have

$$v(\bar{y}_1, \ldots, \bar{y}_s) = 1,$$

for all $v \in \operatorname{Rel}(b_1, \ldots, b_s)$. Thus there is a homomorphism

$$\theta : H \to \langle G, \mathscr{S} \rangle$$

given by: $b_j \mapsto \bar{y}_j$ $(1 \leq j \leq s)$. Since the implications $\mathscr{S}$ hold in $G \times H$, there is a homomorphism

$$\langle G, \mathscr{S} \rangle \to G \times H$$

given by: $a_i \mapsto a_i$ $(1 \leq i \leq r)$ and $\bar{y}_j \mapsto b_j$ $(1 \leq j \leq s)$. So θ is an embedding, and H can be embedded in $\langle G, \mathscr{S}^* \rangle$, as required.

($\Rightarrow$) If K is finitely presented over G, then we can take, as defining relations for K, the relations of G together with a finite set $\mathscr{S}$. By Theorem 4.11,

$$\operatorname{Rel} K \leq_e (\operatorname{Rel} G) \vee \mathscr{S} \equiv_e \operatorname{Rel} G,$$

since $\mathscr{S}$ is finite. If H can be embedded in K, then by Corollary 4.30, $\operatorname{Rel} H \leq_e \operatorname{Rel} K$, and so $\operatorname{Rel} H \leq_e \operatorname{Rel} G$. □

The rest of this chapter is devoted to proving that—although for any given finitely generated group G there exists a group that is universally finitely presented over G—if G is non-trivial, then there is no group K that is finitely presented over G and in which every finite soluble set of equations over G is soluble.

Theorem 6.3 *If G is any finitely generated group, then there exists a group K that is finitely presented over G and that is such that every group finitely presented over G can be embedded in K.*

Proof Let $G = \langle g_1, \ldots, g_k \rangle$. We can effectively enumerate the set of all finite sets $\mathscr{S}$ of equations over G. Let $\mathscr{S}_n$ be the nth such set, and let $x_1, \ldots, x_m$ be the variables that occur in $\mathscr{S}_n$. Let E_n be the free group freely generated by these variables, and, for $w \in G * E_n$, write $\bar{w}$ for wN, where $\langle G, \mathcal{S}_n \rangle = (G * E_n)/N$. As in Lemma 5.1, let

$$\mathrm{e}_i(y, z) = y^{-1}z^{-1}yz^{-i}yz^{-1}y^{-1}z^iy^{-1}zyz^{-i}yzy^{-1}z^i.$$

For $n \in \mathbb{N}$, choose new variables y_n and z_n, and define sets R_n and S_n by

$$R_n = \{r(\mathrm{e}_1(y_n, z_n), \ldots, \mathrm{e}_k(y_n, z_n)) \mid r(\boldsymbol{x}) \in \mathrm{Rel}\,(g_1, \ldots, g_k)\}$$
$$S_n = \{s(\mathrm{e}_1(y_n, z_n), \ldots, \mathrm{e}_k(y_n, z_n)) \mid (s(\boldsymbol{x}) = 1) \in \mathcal{S}_n\}.$$

Then let H_n be the group, generated by two elements a_n and b_n, that has $R_n \cup S_n$ as a set of defining relations (with respect to these generators).

As in Lemma 5.1, and the note below it, we have that the rules

$$\bar{g}_i \mapsto \mathrm{e}_i(a_n, b_n) \quad (1 \leq i \leq k), \qquad \bar{x}_j \mapsto \mathrm{e}_{k+j}(a_n, b_n) \quad (1 \leq j \leq m)$$

define an embedding of $\langle G, \mathcal{S}_n \rangle$ into H_n. Form the free product, H^*, of all the H_n, and embed it in a group $H = \langle a, b \rangle$, as in Lemma 5.1, by the rules $a_n \mapsto \mathrm{e}_{2n}(a, b)$, and $b_n \mapsto \mathrm{e}_{2n+1}(a, b)$ $(n \in \mathbb{N})$. So $\langle G, \mathcal{S}_n \rangle$ is embedded in H $(n \in \mathbb{N})$. Let

$$T = \{t(\mathrm{e}_{2n}(y, z), \mathrm{e}_{2n+1}(y, z)) \mid t(y, z) \in R_n\},$$
$$S = \{v(\mathrm{e}_{2n}(y, z), \mathrm{e}_{2n+1}(y, z)) \mid v(y, z) \in S_n\}.$$

Then we can choose H to have defining relations $T \cup S$. Since the function $i \mapsto \mathrm{e}_i(y, z)$ $(i \in \mathbb{N})$ is recursive, and since $\{\mathcal{S}_n \mid n \in \mathbb{N}\}$ is recursively enumerable, S is recursively enumerable, and $T \cup S \leq_{\mathrm{e}} T$. By Theorem 4.11, $\mathrm{Rel}\, H \leq_{\mathrm{e}} T \cup S$. Suppose that

$$\mathrm{Rel}\,(g_1, \ldots, g_k) \subseteq W = \mathrm{W}(x_1', \ldots, x_k').$$

For each word $w \in W$, and for each integer n, let

$$w^{(n)}(y, z) = w(\mathrm{e}_1(\mathrm{e}_{2n}(y, z), \mathrm{e}_{2n+1}(y, z)), \ldots, \mathrm{e}_k(\mathrm{e}_{2n}(y, z), \mathrm{e}_{2n+1}(y, z))).$$

Then the set $U = \{(w^{(n)}(y, z), \{w\}) \mid w \in W, \ n \in \mathbb{N}\}$ is recursively enumerable. Now, $w^{(n)}(y, z) \in T$ if and only if

$$w(\mathrm{e}_1(y, z), \ldots, \mathrm{e}_k(y, z)) \in R_n,$$

which is if and only if $w \in R$. So

$$T \leq_{\mathrm{e}} \mathrm{Rel}\,(g_1, \ldots, g_k) \text{ via } U,$$

and thus $\mathrm{Rel}\, H \leq_{\mathrm{e}} \mathrm{Rel}\, G$. Then, by Theorem 6.2, we see that H can be embedded in a group K that is finitely presented over G, and each $\langle G, \mathcal{S}_n \rangle$ is embedded in K, as required. □

Corollary 6.4 *For any $X \subseteq \mathbb{N}$, the existentially closed group M_X contains a relatively universal finitely generated subgroup K* (i.e. $\mathrm{Sk}\, K = \mathrm{Sk}\, \mathrm{M}_X$).

Proof $\mathrm{Sk}\,\mathrm{M}_X = \{H \mid H$ finitely generated, and $\mathrm{Rel}\,H \leqslant_{\mathrm{e}} X\}$, and we can choose $G \subseteq \mathrm{M}_X$ such that $\mathrm{Rel}\,G \equiv_{\mathrm{e}} X$. Take K as in Theorem 6.3; then $\mathrm{Rel}\,K \equiv_{\mathrm{e}} \mathrm{Rel}\,G$ so that $K \in \mathrm{Sk}\,\mathrm{M}_X$, and $\mathrm{Sk}\,K \subseteq \mathrm{Sk}\,\mathrm{M}_X$. If $\mathrm{Rel}\,H \leqslant_{\mathrm{e}} \mathrm{Rel}\,G$ then $H \subseteq K_0$, where K_0 is finitely presented over G, by Theorem 6.2. By Theorem 6.3, $K_0 \subseteq K$ and so $H \in \mathrm{Sk}\,K$. Thus, $\mathrm{Sk}\,K = \mathrm{Sk}\,\mathrm{M}_X$. □

Let $\mathcal{S}_n$ be the nth set of equations over G, as above. We can construct a map $\varphi_n : G \to K$ that is the composition of the natural map $G \to \langle G, \mathcal{S}_n \rangle$ and the map $\langle G, \mathcal{S}_n \rangle \to K$, which is given by the construction in the proof of Theorem 6.3. We write $K = \langle G, \mathcal{T} \rangle$, where $\mathcal{T}$ is some finite set of equations over G. Then G is naturally embedded in $\langle G, \mathcal{T} \rangle$; but φ_n cannot be an embedding if $\mathcal{S}_n$ is not soluble over G, i.e. the map $\langle G, \mathcal{S}_n \rangle \to \langle G, \mathcal{T} \rangle$, constructed above, cannot always be G-preserving.

We now consider the question of whether Theorem 6.3 can be strengthened by insisting that we can find K such that, whenever $\mathcal{S}_n$ is soluble over G, some homomorphism $\langle G, \mathcal{S}_n \rangle \to K$ is G-preserving. In other words, we ask whether there exists a group K that is finitely presented over G and is such that, for all finite sets $\mathcal{S}$ of equations, if $\mathcal{S}$ is soluble over G then $\mathcal{S}$ is soluble in K. This would require, for any finite soluble set $\mathcal{S}$, the existence of an embedding $\psi : \langle G, \mathcal{S} \rangle \to K$ such that the diagram

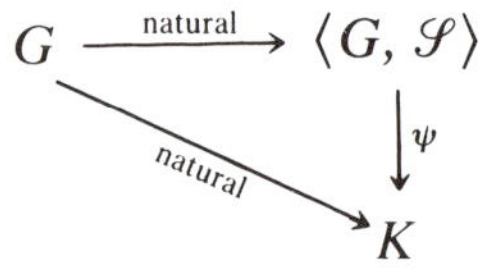

can be made to commute.

If G_1 and G_2 are two copies of G in K, then G_1 and G_2 are conjugate in some HNN-extension K^* of K. Since G is finitely generated, K^* is finitely presented over K, and hence over G. Thus we can ensure, for one particular $\mathcal{S}_n$, and indeed for finitely many such sets, that the map constructed in Theorem 6.3 is G-preserving. But, unless $G = 1$, this cannot be done for all $\mathcal{S}_n$ at once. We take the first step towards showing this, by proving the following theorem.

Theorem 6.5 *Let $\{\mathcal{S}_i \mid i \in \mathbb{N}\}$ be a recursively enumerable collection of finite sets of equations, defined over the finitely generated group G.*

(i) *If $\{i \mid \mathcal{S}_i$ is soluble over $G\} = X \leqslant_{\mathrm{e}} \mathrm{Rel}\,G$, then there exists a group L that is finitely presented over G and is such that $\mathcal{S}_i$ is soluble in L, for all $i \in X$, i.e. $X = \{i \mid \mathcal{S}_i$ is soluble in $L\}$.*

(ii) *If L is a group that is finitely presented over G, and if $X = \{i \mid \mathcal{S}_i$ is soluble in $L\}$, then $X \leqslant_{\mathrm{e}} \mathrm{Rel}\,G$.*

Proof (i) We need to construct a group L that is finitely presented over G, and embeddings

$$\varphi_i : \langle G, \mathcal{S}_i \rangle \to L \quad (i \in X)$$

such that the diagrams

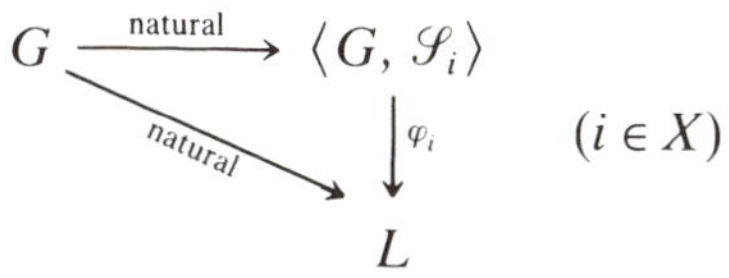

all commute.

Let K be the group constructed in Theorem 6.3 which embeds $\langle G, \mathcal{S}_n \rangle$ for all $n \in \mathbb{N}$. There are natural embeddings

$$\alpha : G \to K, \qquad \alpha_i : G \to \langle G, \mathcal{S}_i \rangle \quad (i \in X),$$

and there are also embeddings

$$\theta_i : \langle G, \mathcal{S}_i \rangle \to K \quad (i \in X).$$

Construct the repeated HNN-extension

$$K_1 = \langle K, t_0, t_1, \ldots \mid t_i^{-1}(g\alpha)t_i = g\alpha_i\theta_i \ (i \in X) \rangle$$

of K by t_n $(n \in \mathbb{N})$ (here t_i conjugates $G\alpha$ to $G\alpha_i\theta_i$ if $i \in X$, and $\{1\}$ to $\{1\}$ if $i \notin X$). So we have the diagrams of Fig. 6.1,

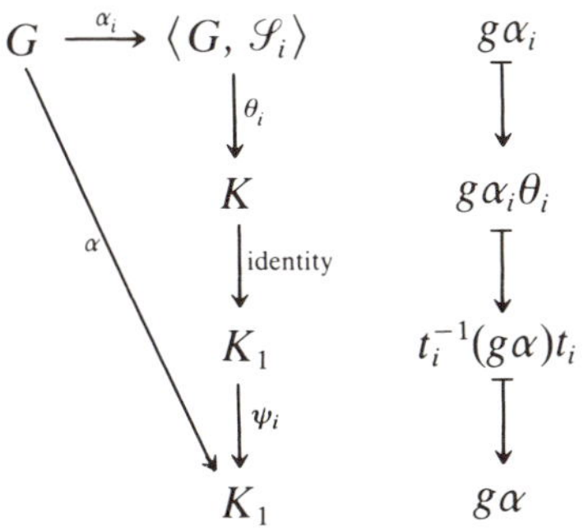

Figure 6.1

for $i \in X$, where $\psi_i \in \operatorname{Aut} K_1$ is conjugation by t_i^{-1}. These diagrams are clearly commutative, and are of the required type except that K_1 need not be finitely presented over G.

Using Lemma 5.1, we can embed K_1 in a group $K_2 = \langle K, x, y \rangle$ under the embedding

$$\theta : K_1 \to K_2,$$

where $\theta\,|_K$ is the identity, and

$$\theta(t_n) = \mathrm{e}_n(x, y) \quad (n \in \mathbb{N}).$$

Then we have the diagrams of Fig. 6.2, for $i \in X$.

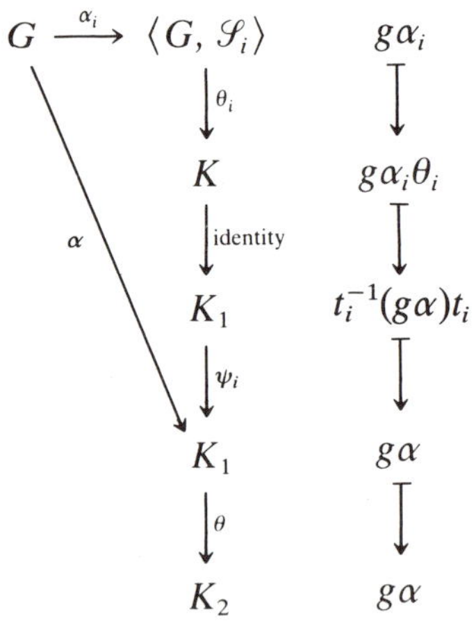

Figure 6.2

As defining relations for K_2, we can take the relations of K together with the relations

$$\{\mathrm{e}_i^{-1}(g\alpha)\mathrm{e}_i = g\alpha_i\theta_i \mid i \in X\},$$

where $\mathrm{e}_i = \mathrm{e}_i(x, y)$. So we have

$$\mathrm{Rel}\, K_2 \leqslant_{\mathrm{e}} (\mathrm{Rel}\, K) \vee X.$$

By Theorem 6.2,

$$\mathrm{Rel}\, K \leqslant_{\mathrm{e}} \mathrm{Rel}\, G,$$

and, by assumption,

$$X \leqslant_{\mathrm{e}} \mathrm{Rel}\, G.$$

So we have that

$$\mathrm{Rel}\, K_2 \leqslant_{\mathrm{e}} \mathrm{Rel}\, G.$$

Thus, by Theorem 6.2, there is a group L_1 that is finitely presented over G, and an embedding

$$\varphi : K_2 \to L_1.$$

Let $\gamma : G \to L_1$ be the natural embedding. We can form the HNN-extension

$$L = \langle L_1, s \mid s(g\gamma)s^{-1} = g\alpha\varphi \ \ (g \in G) \rangle.$$

Since G is finitely generated, L is finitely presented over G. Let $\psi_s \in \mathrm{Aut}\, L$ be conjugation by s. Then we have the commutative diagrams of Fig. 6.3 ($i \in X$).

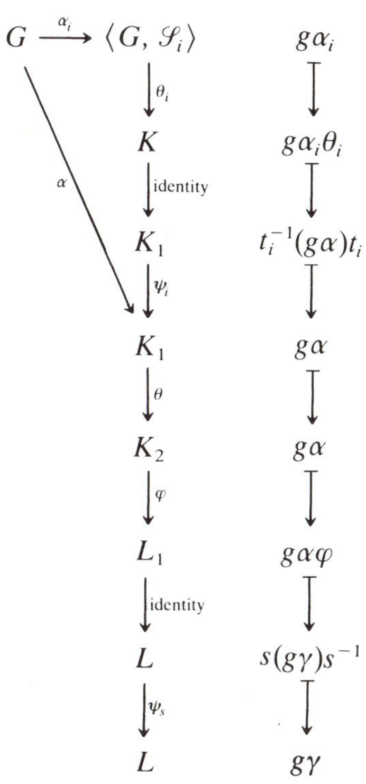

Figure 6.3

So, putting $\varphi_i = \theta_i \psi_i \theta \varphi \psi_s$, we get commutative diagrams

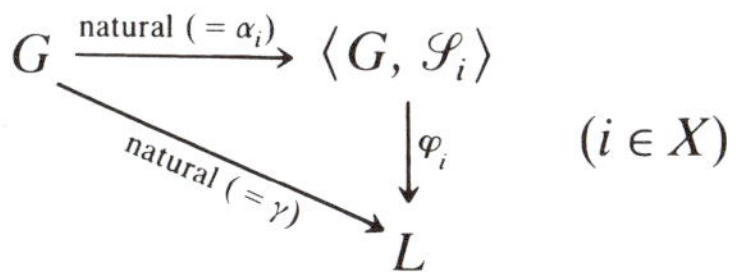

as required.

(ii) First, let $\mathcal{S}$ be any finite set of equations over G. We search for solutions of $\mathcal{S}$ in L, while keeping track of the constructive aspect of the search.

Let

$$\{a_1, \dots, a_n\} \quad \text{and} \quad \{a_1, \dots, a_n, b_1, \dots, b_m\}$$

be sets of generators for G and L, respectively. Let

$$\{z_1, \dots, z_n, y_1, \dots, y_m\}$$

be a corresponding set of distinct variables. We shall be considering sets of words in these variables and their inverses.

We can choose a finite set R of relations of L, which, together with the relations of G, form a set of defining relations for L. Let $\{x_1, \dots, x_r\}$ be the variables occurring in $\mathscr{S}$. We consider r-tuples

$$(w_1, \dots, w_r)$$

of words in $\mathrm{W}(z_1, \dots, z_n, y_1, \dots, y_m)$.

Let $u_1 = 1, u_2 = 1, \dots, u_s = 1$ be the equations in $\mathscr{S}$, and let $u_i(\boldsymbol{w})$ denote the result of substituting w_j for x_j in $u_i(\boldsymbol{x})$ $(1 \leqslant i \leqslant s,\ 1 \leqslant j \leqslant r)$. So $u_i(\boldsymbol{w}) = 1$ is an equation over G in the variables $z_1, \dots, z_n, y_1, \dots, y_m$, and

$$\big(w_1(\boldsymbol{a}, \boldsymbol{b}), \dots, w_r(\boldsymbol{a}, \boldsymbol{b})\big)$$

is a solution of $\mathscr{S}$ in L if

$$\{u_1(\boldsymbol{w}), \dots, u_s(\boldsymbol{w})\} \subseteq \mathrm{Rel}\,(\boldsymbol{a}, \boldsymbol{b}).$$

In this case we can find a finite set E, where

$$E \subseteq \mathrm{Rel}\,(a_1, \dots, a_n),$$

such that $u_i(\boldsymbol{w}) = 1$ is a consequence of $E \cup R$, for $1 \leqslant i \leqslant s$.

For each finite set $A \subseteq \mathrm{W}(z_1, \dots, z_n)$, we can recursively construct the set of all consequences of $A \cup R$. We can effectively enumerate the r-tuples $(w_1, \dots, w_r)$ in $\mathrm{W}(z_1, \dots, y_m)$, and so we can list the s-tuples $\big(u_1(\boldsymbol{w}), \dots, u_s(\boldsymbol{w})\big)$, for each $(w_1, \dots, w_r)$. Periodically comparing the lists of consequences of the sets $A \cup R$ with the list of s-tuples $\big(u_1(\boldsymbol{w}), \dots, u_s(\boldsymbol{w})\big)$, we see that the set of $(s+1)$-tuples

$$U(\mathscr{S}) = \big\{\big(u_1(\boldsymbol{w}), \dots, u_s(\boldsymbol{w}), A\big) \;\big|\; u_i(\boldsymbol{w}) = 1 \text{ is a consequence of } A \cup R, \text{ for } 1 \leqslant i \leqslant s\big\}$$

is recursively enumerable.

Now, for the recursively enumerable set $\{\mathscr{S}_i \mid i \in \mathbb{N}\}$ under consideration, the set

$$V = \big\{(i, A) \;\big|\; (\exists w_1, \dots, w_{r_i})\big[\big(u_{i_1}(\boldsymbol{w}), \dots, u_{i_k}(\boldsymbol{w}), A\big) \in U(\mathscr{S}_i)\big]\big\}$$

is recursively enumerable. Then $X = \{i \mid \mathscr{S}_i \text{ is soluble in } L\}$ is the set of all $i \in \mathbb{N}$ such that $(i, E) \in V$, for some

$$E \subseteq \mathrm{Rel}\,(a_1, \dots, a_n)$$

So $X \leqslant_{\mathrm{e}} \mathrm{Rel}\, G$. □

Remark Let the set $\{\mathscr{S}_i \mid i \in \mathbb{N}\}$ and the group G be as in Theorem 6.5. Then the following two statements are equivalent.

(a) There exists a group L that is finitely presented over G and such that $\mathscr{S}_i$ is soluble in L whenever it is soluble over G.

(b) If $J = \{i \mid \mathscr{S}_i \text{ is soluble over } G\}$, then $J \leqslant_{\mathrm{e}} \mathrm{Rel}\, G$.

For, Theorem 6.5(i) shows that (b)$\Rightarrow$(a), and Theorem 6.5(ii) shows that (a)$\Rightarrow$(b).

Lemma 6.6 *If $G \neq \{1\}$ and H are finitely generated groups with $\mathrm{Rel}\, H \leqslant_e \mathrm{Rel}\, G$, then we can choose a recursively enumerable set $\{\mathscr{S}_i \mid i \in \mathbb{N}\}$ of finite sets of equations over G, such that $J = \{i \mid \mathscr{S}_i$ is soluble over $G\} \equiv_1 \mathrm{Nonrel}\, H = \mathbb{N} \backslash \mathrm{Rel}\, H$.*

Proof To say that the set $\{\mathscr{S}_i \mid i \in \mathbb{N}\}$ is recursively enumerable means that there is a machine M which, for input i, produces output $w_1, w_2, \ldots, w_{n_i}$ such that

$$\{w_1 = 1, \ldots, w_{n_i} = 1\} = \mathscr{S}_i.$$

Suppose that $H = \langle h_1, \ldots, h_t \rangle$. There is a recursive bijection θ from $\mathbb{N}$ to the set of all words on the generators $h_1, \ldots, h_t$ and their inverses, (here we mean *words*, so that $h_1 h_1^{-1}$ is a different word from $h_2 h_2^{-1}$, etc.).

Since $\mathrm{Rel}\, H \leqslant_e \mathrm{Rel}\, G$, by The Relative-Subgroup Theorem, there is a finite set $\mathscr{S}$ of equations which are defined and soluble over G, and such that there is an embedding

$$\varphi : H \to \langle G, \mathscr{S} \rangle.$$

Let $x_1, \ldots, x_s$ be the variables in $\mathscr{S}$, and choose new variables y and z. Choose $g \in G \backslash \{1\}$, let

$$u_i = \theta(i) \quad (i \in \mathbb{N}),$$

and let

$$\mathscr{S}_i = \{y^{-1} u_i y z^{-1} u_i z = g\} \cup \mathscr{S}.$$

Then the set $\{\mathscr{S}_i \mid i \in \mathbb{N}\}$ is recursively enumerable.

Suppose that $h_i \varphi = w_i(\mathbf{g}, \mathbf{x})$ $(1 \leqslant i \leqslant t)$. If $u_i \in \mathrm{Rel}\,(h_1, \ldots, h_t)$, then $u_i = 1$ in H, so

$$u_i(h_1 \varphi, \ldots, h_t \varphi) = u_i(w_1, \ldots, w_t) = 1$$

is a consequence of $\mathscr{S} \cup \mathrm{Rel}\, G$. If

$$\bar{x}_1, \ldots, \bar{x}_s, \bar{y}, \bar{z}$$

are solutions for $\mathscr{S}_i$, then, since $\bar{x}_1, \ldots, \bar{x}_s$ are solutions for $\mathscr{S}$, we have

$$\bar{u}_i = u_i\big(w_1(\mathbf{g}, \bar{x}_1, \ldots, \bar{x}_s), \ldots, w_t(\mathbf{g}, \bar{x}_1, \ldots, \bar{x}_s)\big) = 1.$$

Thus $g = \bar{y}^{-1} \bar{u}_i \bar{y} \bar{z}^{-1} \bar{u}_i \bar{z} = 1$, which is a contradiction. So if $u_i \notin \mathrm{Nonrel}\,(h_1, \ldots, h_t)$, then $\mathscr{S}_i$ is not soluble over G.

Conversely, if $u_i \in \mathrm{Nonrel}\,(h_1, \ldots, h_t)$, then $u_i \neq 1$ in H, and so $u_i \varphi \neq 1$ and $u_i(w_1, \ldots, w_t) = 1$ cannot be a consequence of $\mathscr{S} \cup \mathrm{Rel}\, G$. Thus, whenever we obtain $\bar{u}_i$ from u_i by substituting solutions of $\mathscr{S}$, we must have $\bar{u}_i \neq 1$. Then

$$y^{-1} \bar{u}_i y z^{-1} \bar{u}_i z = g$$

is an equation over $\langle G, \mathscr{S} \rangle$, which, by Theorem 1.5, is soluble over $\langle G, \mathscr{S} \rangle$. Thus $\mathscr{S}_i$ is soluble over G.

So we have shown that

$$\text{Nonrel}\,(h_1, \ldots, h_t) = \{u_i \mid \mathscr{S}_i \text{ is soluble over } G\}.$$

Since

$$\{u_i \mid \mathscr{S}_i \text{ is soluble over } G\} \equiv_1 \{i \mid \mathscr{S}_i \text{ is soluble over } G\} \text{ via } \theta,$$

we have that

$$\{i \mid \mathscr{S}_i \text{ is soluble over } G\} \equiv_1 \text{Nonrel}\, H. \qquad \square$$

Essentially re-writing Lemma 4.18, we get the following result.

Lemma 6.7 *For any subset Z of $\mathbb{N}$ there exists a finitely generated group H such that* $\text{Rel}\, H \leqslant_e Z$*, but* $\text{Nonrel}\, H \nleqslant_e Z$.

Proof Choose X, as in Lemma 4.18, such that $X \leqslant_e Z$, but $\mathbb{N}\backslash X \nleqslant_e Z$. By Corollary 4.24, we can choose a finitely generated group, H $(= G_X)$, such that

$$X \equiv^+ \text{Rel}\, H.$$

For any set $Y \subseteq \mathbb{N}$ such that $Y \leqslant^+ X$ via f, we have $\mathbb{N}\backslash Y \leqslant^+ \mathbb{N}\backslash X$. To see this, define

$$g : \mathbb{N} \to \mathscr{P}_f(\mathbb{N}) \cup \{\infty\}$$

by

$$g(n) = \begin{cases} \varnothing & \text{if } f(n) \nsubseteq X \quad \text{or} \quad \text{if } f(n) = \infty, \\ \infty & \text{if } f(n) \subseteq X. \end{cases}$$

Then

$$\begin{aligned} n \in \mathbb{N}\backslash Y \quad &\Leftrightarrow \quad f(n) \nsubseteq X \quad \text{or} \quad f(n) = \infty \\ &\Leftrightarrow \quad g(n) = \varnothing \\ &\Leftrightarrow \quad g(n) \neq \infty \quad \text{and} \quad g(n) \subseteq \mathbb{N}\backslash X. \end{aligned}$$

So we have $\text{Nonrel}\, H \equiv^+ \mathbb{N}\backslash X$. Since $Y \leqslant^+ X$ implies that $Y \leqslant_e X$, we get $\text{Rel}\, H \leqslant_e Z$, and

$$\text{Nonrel}\, H \equiv_e \mathbb{N}\backslash X \nleqslant_e Z. \qquad \square$$

From Theorem 6.5, together with Lemma 6.6 and Lemma 6.7, we get the following result.

Theorem 6.8 *If G is a non-trivial group, there does not exist a group K that is finitely presented over G and that is such that every finite set of equations soluble over G is soluble in K.*

Proof Suppose such a group K exists. By Lemma 6.7, we can find a finitely generated group H such that

$$\mathrm{Rel}\, H \leqslant_{\mathrm{e}} \mathrm{Rel}\, G \quad \text{and} \quad \mathrm{Nonrel}\, H \not\leqslant_{\mathrm{e}} \mathrm{Rel}\, G.$$

By Lemma 6.6, we can choose a recursively enumerable set $\{\mathcal{S}_i \mid i \in \mathbb{N}\}$ of finite sets of equations, which are defined over G, such that

$$J = \{i \mid \mathcal{S}_i \text{ is soluble over } G\} \equiv_{\mathrm{e}} \mathrm{Nonrel}\, H.$$

But,

$$J = \{i \mid \mathcal{S}_i \text{ is soluble in } K\},$$

so, by Theorem 6.5(ii), $J \leqslant_{\mathrm{e}} \mathrm{Rel}\, G$. Hence

$$\mathrm{Nonrel}\, H \leqslant_{\mathrm{e}} \mathrm{Rel}\, G,$$

which is a contradiction. □

As a corollary of this, we get the following results.

Theorem 6.9 *If H and G are finitely generated groups, and if M is an existentially closed group with*

$$G < M < H,$$

then H is not finitely presented over G.

Corollary 6.10 *No existentially closed group is embeddable in a finitely presented group.*

Corollary 6.11 *No existentially closed group is embeddable in a finitely generated subgroup of itself, i.e. $G < M < K$ implies that G is not isomorphic to K.*

Proof If $G < M < K$ and $K \cong G$, then the HNN-extension $\langle K, t \mid t^{-1}Kt = G\rangle$ is finitely presented over K, since K is finitely generated, and hence is finitely presented over G, since G and K are conjugate. □

We conclude this chapter with a result, in the form of an exercise, whose proof is left to the reader.

Exercise Construct a finitely generated group G that has a subgroup G_0 such that G is isomorphic to G_0 but G is not finitely presented over G_0.

7
Games

The object of this chapter is to prove the converses of various theorems that we have proved above, and to show that there exists a set X, of cardinality $2^{\aleph_0}$, of countable existentially closed groups such that the intersection of the skeletons of any two distinct elements of X consists precisely of the groups with solvable word problem.

We consider a sequence $g_0, g_1, g_2, \ldots$ of potential group generators. We write W for the set of all words on $g_0^{\pm 1}, g_1^{\pm 1}, \ldots$, and F for the free group freely generated by the g_i.

A *position* $P = (U, V)$ is a consistent pair of subsets of W. (Recall from Chapter 4, that a pair (U, V) is consistent if $\langle U \rangle^F \cap V = \emptyset$.)

A *move* is a pair of positions, $P_1 = (U_1, V_1)$ and $P_2 = (U_2, V_2)$, such that $U_1 \subseteq U_2$ and $V_1 \subseteq V_2$.

A *game* is a sequence of positions, $P_0, P_1, P_2, \ldots$, such that, for each $i \in \mathbb{N}$, the pair (P_i, P_{i+1}) is a move.

A game is said to *yield* the group

$$Y = \left\langle g_0, g_1, g_2, \ldots \;\middle|\; u = 1; \quad u \in \bigcup_{i=0}^{\infty} U_i \right\rangle,$$

and the non-relation $v \neq 1$ holds in Y for all $v \in \bigcup_{i=0}^{\infty} V_i$, since the positions are consistent.

We suppose that our games are played by two players, who take it in turns to make moves.

We do not necessarily want to allow the first player to define the yield Y of the game completely (which he could do by picking U_0 so that Y has the relations he requires and then picking $V_0 = W \backslash \langle U_0 \rangle^F$), so we shall assume that our games are played under some *code of rules*. We will not formally define what we mean by a code of rules; we will merely describe certain examples. But any code of rules will define a set of legal positions, which always includes the empty set. Then it will be allowed, under the rules, to make the move (P, Q) at any suitable point (i.e. any point at which the player is faced with the position P) during the game, provided that Q is one of the legal positions defined by the code of rules.

Examples *The finite code of rules* The finite-legal positions are precisely the finite positions $P = (U, V)$ where $|U| < \infty$ and $|V| < \infty$. Since we are dealing with an infinite set of generators, at no point during a game, being played under the finite rules, can either player completely determine Y.

The stable code of rules For $A \subseteq \mathbb{N}$, let W_A be the set of all words on $\{g_i^{\pm 1} \mid i \in A\}$. Then the stable-legal positions are precisely those positions of the form $P = (U, V)$, where for some finite $A \subseteq \mathbb{N}$, $U \subseteq W_A$ and $V = W_A \backslash U$. Each stable-legal position P defines a finitely generated group Y_A, where Y_A is generated by $\{g_i \mid i \in A\}$ and has U as a set of defining relations. Since (U, V) is consistent, $\langle U \rangle^F \cap V = \varnothing$. So $\langle U \rangle^F \cap W_A = U$. Thus $U = \text{Rel}\, Y_A$ and $V = \text{Nonrel}\, Y_A$. If $P_0, P_1, P_2, \ldots$ is a game played under the stable code of rules, with $P_k = (U_k, V_k)$ and $U_k \subseteq W_{A_k}$, then, since $P_k \subseteq P_{k+1}$, we have that Y_{A_k} is a subgroup of $Y_{A_{k+1}}$, and the yield of the game is $\bigcup_{k=0}^{\infty} Y_{A_k} = Y$. The name 'stable' for this code arises because, once a player has forced a group to be a subgroup of Y, then neither player can change this. Under the finite code of rules, if one player introduces a subgroup during the game, the other player can often reduce this to a proper homomorphic image of itself, at some later stage in the same.

If R is a code of rules, then a property $\mathcal{P}$, of groups, is said to be *R-enforceable* if there is a second-player strategy guaranteeing that any game played under the rules R yields a group that has the property $\mathcal{P}$.

So, every time the first player makes a move, there is a corresponding move that the second player can make to ensure that, at the end of the game, the yield has the property $\mathcal{P}$.

For example, the property

'has an element of order 3'

is finite-enforceable. If the first move is to the position P_0, the second player chooses an integer i such that g_i is not involved in P_0. Then

$$P_1 = (U_0 \cup \{g_i^3\}, \quad V_0 \cup \{g_i\})$$

is a finite-legal position. So the second player makes the move (P_0, P_1).

Of course, the property

'all elements have order 3'

is not finite-enforceable.

We say that a code of rules R is *acceptable* if:

(i) being existentially closed is R-enforceable,

(ii) for any property $\mathcal{P}$ of groups, either $\mathcal{P}$ or (not $\mathcal{P}$) is R-enforceable,

(iii) whenever all the properties in the set $\{\mathcal{P}_i \mid i \in \mathbb{N}\}$ are R-enforceable, then $\bigcap_{i \in \mathbb{N}} \mathcal{P}_i$ is also R-enforceable.

Note We could not necessarily expect (ii) to hold if we were taking the yield to be something other than a group. For example, if games yielded fields $\mathbb{F}$ (say), the first player could ensure that Char $\mathbb{F} = 5$, so there would be no second-player strategy which would force Char $\mathbb{F} = 2$. Similarly, the first player could ensure that Char $\mathbb{F} = 2$, so there would be no second-player strategy that forced Char $\mathbb{F} \neq 2$, either.

The codes of rules that we will consider will satisfy:

(a) If π is a permutation of the set $\{g_i \mid i \in \mathbb{N}\}$ and if P is a legal position, then $P\pi$, the set obtained by substituting $g_i\pi$ for g_i $(i \in \mathbb{N})$ in the words in P, must also be a legal position.

(b) Legal positions must have finite support (i.e. must involve only finitely many of the g_i).

(c) If P is a legal position, if P_0 is either a legal or a finite position, and if $P \cup P_0$ is consistent, then there is a legal position containing $P \cup P_0$. (If $P = (U, V)$ and $Q = (X, Y)$ then

$$P \cup Q = (U \cup X, V \cup Y).)$$

In general, it is hard to tell whether a given code of rules satisfies (i), (ii), and (iii) above. So the following lemma gives us a more practical way of showing that a code of rules is acceptable.

Lemma 7.1 *A code of rules R that satisfies* (*a*), (*b*), *and* (*c*), *above, is acceptable.*

Proof (i) Let Y denote the yield of the game. We show that, under R, we can force Y to be a non-trivial algebraically closed group. If P_0 is the first position in the game, then, by (b), we can choose a g_i that does not occur in P_0. Hence, by (c), there is a legal position P_1 containing $P_0 \cup (\varnothing, \{g_i\})$. So the move (P_0, P_1) ensures that the inequality $g_i \neq 1$ holds in Y. Hence we can enforce $Y \neq 1$. We wish then to show that any finite set of equations over Y that is soluble over Y is soluble in Y. Of course, we cannot identify Y until the game has been played, but any equation over Y is also an equation over the free group F generated by $g_0, g_1, \ldots$. So we shall show that we can force Y to have the property that any finite set $\mathscr{S}$ of equations over F that is soluble over Y is soluble in Y. The set of all finite sets of equations over F is countable, so it is sufficient to show that, in one legal move, we can deal with any one $\mathscr{S}$. Suppose that we are faced with the position $P = (U, V)$ and let

$$S = \{w \mid w = 1 \text{ is an equation in } \mathscr{S}\}.$$

By definition of Y, the sets $\{u = 1 \mid u \in U\}$ and $\{v \neq 1 \mid v \in V\}$, of equations and inequalities, will hold in Y. So, if the pair $P \cup (S, \varnothing)$ is inconsistent, then $\mathcal{S}$ cannot be consistent with the relations and non-relations of Y. So $\mathcal{S}$ is not soluble over Y, and in this case we need do nothing: our second-player move may be (P, P). If $P \cup (S, \emptyset)$ is consistent, the words in P involve only $g_0, \ldots, g_m$ (say). Now replace the variables $x_1, \ldots, x_r$ in $\mathcal{S}$, by $g_{m+1}, \ldots, g_{m+r}$, to obtain a new set $\mathcal{S}_0$. Then $P \cup (S_0, \emptyset)$ is consistent and, since $\mathcal{S}_0$ is finite, there is a legal position Q containing $P \cup (S_0, \emptyset)$, by (c). We take (P, Q) as our second-player move; then $\mathcal{S}$ is soluble in Y, being satisfied by $g_{m+1}, \ldots, g_{m+r}$.

(ii) For every property $\mathcal{P}$, any yield either has or does not have $\mathcal{P}$. So, if there is no second-player strategy which forces Y to have $\mathcal{P}$, there is a first-player strategy which forces Y to have (not $\mathcal{P}$). Let us call this first-player strategy **S**. We show that **S** can be converted into a second-player strategy. Suppose that the first move has been made, to the position P. The strategy **S** produces a first move to the position Q, say. Choose a permutation π of the set $\{g_i \mid i \in \mathbb{N}\}$, such that the supports of P and $Q\pi$ are disjoint. Then P and $Q\pi$ are legal positions, by (a), and $P \cup Q\pi$ is consistent; so, by (c), there exists a legal position T containing $P \cup Q\pi$. Hence there is a game which begins Q, $T\pi^{-1}$. Then **S** produces a move to a position N (say), so that this game proceeds Q, $T\pi^{-1}$, N. Thus, $N\pi$ is a legal position, and the second player makes the move $(P, N\pi)$. If the first player then makes the move $(N\pi, E)$, we may suppose that our alternative game proceeds Q, $T\pi^{-1}$, N, $E\pi^{-1}$. Then **S** produces the move $(E\pi^{-1}, F)$, and we play the move $(E, F\pi)$, in the original game. Thus, if the original game proceeds

$$P_0, P_1, P_2, \ldots$$

there is an alternative game, which proceeds

$$Q_0, Q_1, Q_2, \ldots$$

such that $P_{i-1} = Q_i\pi$, with the first player in the alternative game playing according to **S**. Since $\bigcup_{i\in\mathbb{N}} P_i = \bigcup_{i\in\mathbb{N}} Q_i\pi$, the two games have isomorphic yields, and so we can ensure that Y does not have $\mathcal{P}$.

(iii) Divide the natural numbers $\mathbb{N}$ into infinitely many disjoint infinite sets X_i $(i \in \mathbb{N})$. We suppose that the second player is about to make his jth move in the game, and that $j \in X_i$. So far, the game has proceeded to $P_0, P_1, P_2, \ldots, P_{2j-2}$, and within this game, we have a subgame

$$P_{i_1}, P_{i_1+1}, P_{i_2}, P_{i_2+1}, \ldots, P_{2j-2},$$

with $0 \leqslant i_1 < i_2 < \cdots < 2j-2$, and $\frac{1}{2}i_{k+2} \in X_i$, which is being played using a second-player strategy $\mathbf{S}_i$ to ensure that the yield has the property $\mathcal{P}_i$. For his jth move in the main game, the second player moves to the position P_{2j-1}, where (P_{2j-2}, P_{2j-1}) is the move produced by $\mathbf{S}_i$. Since $P_0 \subseteq P_1 \subseteq \cdots \subseteq P_l$, all the moves $(P_{i_k+1}, P_{i_{k+1}})$ in the subgame are legal. In this way we ensure that Y has all the properties $\mathcal{P}_i$, as required. □

At this point, we define two more codes of rules.

The (A, B) code of rules Here A and B are subsets of $\mathbb{N}$, and a position $P=(U, V)$ is (A, B)-legal if $(U, V) \leqslant^* (A, B)$, and if P has finite support.

The minimal code of rules The minimal-legal positions are precisely those positions having finite support.

It follows from Theorem 7.1 that the finite, stable, (A, B), and minimal codes of rules are all acceptable.

Two codes of rules R_1 and R_2 are said to be *equivalent* if a property $\mathscr{P}$ is R_1-enforceable precisely when it is R_2-enforceable.

1. *The stable code of rules and the minimal code of rules are equivalent.* This is because if $\mathscr{P}$ is, for example, a property which is stable-enforceable using the strategy $\mathbf{S}$, and if $P_0, P_1, P_2, \ldots$ is any game played under the minimal rules, then there is a parallel game $Q_0, Q_1, Q_2, \ldots$, played under the stable rules and using $\mathbf{S}$, that has a yield isomorphic to the yield of the original game.

To show that we can construct such a parallel game, we need to note that any stable-legal position is also a minimal-legal position. Also, if $P=(U, V)$ is a minimal-legal position, then there is a finitely generated group G such that $U \subseteq \text{Rel}\, G$ and $V \subseteq \text{Nonrel}\, G$. So there is a stable-legal position $Q=(\text{Rel}\, G, \text{Nonrel}\, G)$, containing P.

Now suppose that $\mathbf{S}$ is a strategy allowing the second player to force $\mathscr{P}$ to be a property of the yield of a game played under the stable code of rules. Suppose, also, that a minimal game begins with a move to position P_0. Let Q_0 be a stable-legal positon containing P_0. Then there is a stable game beginning with a move to Q_0. So $\mathbf{S}$ produces a second move (Q_0, Q_1). Then Q_1 is also a minimal-legal position and $P_0 \subseteq Q_0 \subseteq Q_1$, so the move (P_0, Q_1) is minimal-legal, and we make this move. If the next move in the minimal game is (Q_1, P_2), then, if Q_2 is a stable-legal position containing P_2, it follows that (Q_1, Q_2) is a stable-legal move. Thus, there is a stable game that begins Q_0, Q_1, Q_2, and $\mathbf{S}$ produces a move (Q_2, Q_3). The second player then makes the move (P_2, Q_3) in the minimal game. Continuing in this way, we see that, if the minimal game proceeds

$$P_0, P_1, P_2, \ldots, P_n, \ldots,$$

then there is a parallel stable game which proceeds

$$Q_0, Q_1, Q_2, \ldots, Q_n, \ldots,$$

with $Q_{2i-1}=P_{2i-1}$ and $P_{2i} \subseteq Q_{2i}$. Now, the yield of the original game is

$$Y=\left\langle g_0, g_1, \ldots \;\middle|\; u=1, \quad u \in \bigcup_{i=0}^{\infty} U_i \right\rangle$$

and the yield of the parallel game is

$$Y_1 = \left\langle g_0, g_1, \dots \;\middle|\; u = 1, \quad u = \bigcup_{i=0}^{\infty} X_i \right\rangle,$$

where $P_i = (U_i, V_i)$ and $Q_i = (X_i, Y_i)$. Since $Q_i \subseteq P_{i+1}$ and $P_i \subseteq Q_{i+1}$, we have that $\bigcup_{i=0}^{\infty} U_i = \bigcup_{i=0}^{\infty} X_i$. So Y is isomorphic to Y_1 and, since Y_1 has been forced to have $\mathcal{P}$, it follows that Y is forced to have $\mathcal{P}$, as required.

Exactly the same type of construction shows that, if $\mathcal{P}$ is a property which is minimal-enforceable, then $\mathcal{P}$ is stable-enforceable. We leave this to the reader to check for himself.

2. *The finite code of rules and the* $(\varnothing, \varnothing)$ *code of rules are equivalent.* To see this, we construct parallel games, as in the first example. However, this case is more complicated because, although any finite-legal position is also a $(\varnothing, \varnothing)$-legal position, there exist infinite $(\varnothing, \varnothing)$-legal positions, which can therefore never be contained in a finite-legal position.

For any position $P = (U, V)$ of any game, we define the set of *consequences* of P to be

$$\bar{P} = \langle U \rangle^F \cup \{z \in W \;|\; \langle U, z \rangle^F \cap V \neq \varnothing\},$$

where F is the free group freely generated by $g_0, g_1, g_2, g_3, \dots$; thus, if $x \in \bar{P}$, then either the equation $x = 1$ or the inequality $x \neq 1$ holds in every group in which the equations $\{u = 1 \;|\; u \in U\}$ and the inequalities $\{v \neq 1 \;|\; v \in V\}$ all hold.

If $P = (U, V)$, we shall write P^+ for $\langle U \rangle^F$ and P^- for the set $\{z \in W \,|\, \langle U, z \rangle^F \cap V \neq \varnothing\}$. So that $\bar{P} = P^+ \cup P^-$.

If $P = (U, V)$ and $Q = (X, Y)$ are positions, we shall write $\bar{P} \leqslant \bar{Q}$ if both $P^+ \subseteq Q^+$ and $P^- \subseteq Q^-$. We are interested in this relationship, because, if $P \cup Q$ is to be a position, it has to be consistent. If $\bar{P} \leqslant \bar{Q}$, then

$$\langle X \cup U \rangle^F \cap \langle Y \cup V \rangle = \langle X \rangle^F \cap (Y \cup V) = \langle X \rangle^F \cap V,$$

since Q is a position and therefore consistent. Now, $V \subseteq P^- \subseteq Q^-$, and, if $v \in V \cap \langle X \rangle^F$, then $\langle X \rangle^F = \langle X, v \rangle^F$; so $\langle X \rangle^F \cap Y = \langle X, v \rangle^F \cap Y \neq \varnothing$ (since $v \in P^- \subseteq Q^-$), which is a contradiction. So, if $\bar{P} \leqslant \bar{Q}$,

$$\langle X \cup U \rangle^F \cap (Y \cup V) = \varnothing$$

and $P \cup Q$ is a position.

We show that, given any $(\varnothing, \varnothing)$-legal position P, there exists a finite legal position Q such that $\bar{P} \leqslant \bar{Q}$.

Let F_n denote the free group freely generated by $g_0, g_1, \dots, g_n$. Suppose that $P = (U, V)$, that P has finite support $g_0, g_1, \dots, g_m$, and that $(U, V) \leqslant^* (\varnothing, \varnothing)$. Then both U and V are recursively enumerable. We construct a finite position Q in two stages. First we construct a position $Q' = (U_1, B)$, with B finite and $U \subseteq U_1$. Then we construct $Q = (A, B)$ such that $\bar{P} \leqslant \bar{Q}$.

If $V = \varnothing$, choose $B = \varnothing$. If $V \neq \varnothing$, choose $v \in V$ and let $B = \{v\}$. If $B = \varnothing$, let $J = \varnothing$; if $B \neq \varnothing$, let J be a recursively enumerable subset of $\mathbb{N}\backslash\{0, 1, 2, \ldots, m+2\}$, which can be used to index V. So,

$$V = \{v_j \mid j \in J\}.$$

Let

$$T = \{v^{-1}s_j^{-1}v_js_jt_j^{-1}v_jt_j \mid j \in J\},$$

where s_j and t_j $(j \in J)$ are new variables, and let

$$w_j(g_0, \ldots, g_m, s_j, t_j) = v^{-1}s_j^{-1}v_js_jt_j^{-1}v_jt_j.$$

By Lemma 5.1, the group

$$H_1 = \langle g_0, \ldots, g_m, s_j, t_j \mid u = 1, \quad w_j = 1; \quad u \in U, \quad j \in J\rangle$$

can be embedded in the finitely presented group

$$H = \langle g_{m+1}, g_{m+2} \mid u(e_0, \ldots, e_m) = 1, \quad w_j(e_0, \ldots, e_m, e_{2j-1}, e_{2j}) = 1 \quad (u \in U, \quad j \in J)\rangle.$$

where

$$e_i = e_i(g_{m+1}, g_{m+2}) = [g_{m+1}^{-1}g_{m+2}g_{m+1}, g_{m+2}^{-i}g_{m+1}g_{m+2}g_{m+1}^{-1}g_{m+2}^{i}].$$

Let

$$U_1 = U \cup \{w_j(g_0, \ldots, g_m, e_{2j-1}, e_{2j}) \mid j \in J\} \cup \{e_ig_i^{-1} \mid 0 \leqslant i \leqslant m\},$$

so that H also has presentation

$$\langle g_0, \ldots, g_{m+2} \mid u_1 = 1 \quad (u_1 \in U_1)\rangle.$$

If there is a group $\langle g_0, \ldots, g_{m+2}\rangle$ in which

$$\{u_1 = 1 \mid u_1 \in U_1\} \cup \{v_j \neq 1 \mid j \in J\}$$

all hold, then (U_1, V) is a consistent pair. The reason for this is as follows.

Since $u' = 1$ in $\langle g_0, \ldots, g_{m+2}\rangle$ for all $u' \in \langle U_1\rangle^{F_{m+2}}$, we have

$$\langle U_1\rangle^{F_{m+2}} \cap V = \varnothing$$

(remember, F_{m+2} is the free group freely generated by $g_0, ,,, , g_{m+2}$). But, since $V \subseteq F_{m+2}$, by the Normal-Form Theorem for free products we must therefore have

$$\langle U_i\rangle^F \cap V = \varnothing.$$

Since (U, V) is consistent, there is a group L in which the equations and inequalities

$$\{u=1 \mid u\in U\}\cup\{v_j\neq 1 \mid j\in J\}$$

all hold. Then $v\neq 1$ holds in L, and so, as in the proof of Lemma 1.7, there is an HNN-extension L_1 of L in which the equations

$$\{w_j=1 \mid j\in J\}$$

also hold. Since L_1 is a homomorphic image of H_1, we have that the inequalities

$$\{v_j\neq 1 \mid j\in J\}$$

must hold in H_1, and hence in H. Thus, (U_1, V) and hence (U_1, B) are consistent pairs.

Now let

$$Q_1=(U_1, B).$$

Then $U\subseteq U_1$, and so $\langle U\rangle^F\subseteq\langle U_1\rangle^F$. If $z\in P^-$, then

$$\langle U, z\rangle^F\cap V\neq\varnothing.$$

Suppose that

$$v_j\in\langle U, z\rangle^F\subseteq\langle U_1, z\rangle^F.$$

Since

$$v^{-1}e_{2j-1}^{-1}v_je_{2j-1}e_{2j}^{-1}v_je_{2j}\in U_1$$

we must have $v\in\langle U_1, z\rangle^F$, and hence $z\in Q_1^{-1}$. So

$$P\leqslant Q_1.$$

Now, U_1 is recursively enumerable, so H can be embedded in a finitely generated group K. We can choose any group isomorphic to K, so we may suppose that H is embedded in

$$K=\langle g_{m+3},\ldots,g_n \mid A_1=\{1\}\rangle,$$

for some finite set A_1, and for some $n\geqslant m+3$. Suppose that this embedding carries

$$g_i \quad \text{to} \quad r_i(g_{m+3},\ldots,g_n) \quad (0\leqslant i\leqslant m+2),$$

and let

$$A=A_1\cup\{r_ig_i^{-1} \mid 0\leqslant i\leqslant m+2\}.$$

Then we have

$$K=\langle g_0,\ldots,g_n \mid A=\{1\}\rangle,$$

and $H\subseteq K$.

Since $v \neq 1$ holds in H, it holds in K, and so (A, B) $\big(=(A, \{v\})\big)$ is consistent.

Let

$$Q = (A, B).$$

Then, since $U_1 \subseteq \langle A \rangle^{F_n}$, we have $Q_1^+ \subseteq Q^+$, and since

$$\langle U_1, z \rangle^F \subseteq \big\langle \langle U_1 \rangle^F, z \big\rangle^F \subseteq \langle A, z \rangle^F,$$

we have $Q_1^- \subseteq Q^{-1}$. So

$$\bar{Q}_1 \leqslant \bar{Q}.$$

Since this inequality is clearly transitive, we have

$$\bar{P} \leqslant \bar{Q}.$$

Now suppose that the property $\mathscr{P}$ is finite-enforceable, using the strategy $\mathbf{S}$, and that we are playing a game under the $(\emptyset, \emptyset)$ code of rules. (We need to note here that it is easily seen that if P and Q are pairs such that $P \subseteq Q$, then $\bar{P} \leqslant \bar{Q}$.) Now, let the first move of our game be to the position P_0. Choose a finite position Q_0, as above, so that $\bar{P}_0 \leqslant \bar{Q}_0$. Then $\mathbf{S}$ determines a second move (Q_0, Q_1), in a finite game which had begun with a move to Q_0. We cannot just make the move (P_0, Q_1) in the original game, because we may not have $P_0 \subseteq Q_1$. But the position

$$P_1 = P_0 \cup Q_1$$

is consistent, since $\bar{P}_0 \leqslant \bar{Q}_1$, and it is $(\emptyset, \emptyset)$-legal, since Q_1 is finite. Thus we make the move (P_0, P_1). If the next move in the $(\emptyset, \emptyset)$-game is (P_1, P_2), we choose a finite position Q_2 such that $\bar{P}_2 \leqslant \bar{Q}_2$. Since $Q_1 \subseteq P_1 \subseteq P_2$, we have

$$\bar{Q}_1 \leqslant \bar{P}_1 \leqslant \bar{P}_2 \leqslant \bar{Q}_2,$$

and so $Q_1 \cup Q_2$ is a finite-legal position. So we may choose Q_2 so that $Q_1 \subseteq Q_2$, and still have $\bar{P}_2 \leqslant \bar{Q}_2$. Then there is a finite-game which begins Q_0, Q_1, Q_2. If $\mathbf{S}$ determines the move (Q_2, Q_3) then $P_3 = Q_2 \cup P_2$ is a $(\emptyset, \emptyset)$-legal position, for $\bar{P}_2 \leqslant \bar{Q}_2 \leqslant \bar{Q}_3$. Thus we make the move (P_2, P_3). In this way we see that, if the original game proceeds

$$P_0, P_1, P_2, \ldots, P_n, \ldots,$$

then there is a parallel finite-game which proceeds

$$Q_0, Q_1, Q_2, \ldots, Q_n, \ldots,$$

with $\bar{P}_i \leqslant \bar{Q}_{i+1}$ and $\bar{Q}_i \leqslant \bar{P}_{i+1}$ $(i \in \mathbb{N})$. If the relation $z = 1$ holds in the yield of one of these games, then

$$z \in \bigcup_{i=0}^{\infty} \bar{U}_i = \bigcup_{i=0}^{\infty} \bar{X}_i,$$

(notice, $\langle\bigcup_{i=0}^{\infty} U_i\rangle^F = \bigcup_{i=0}^{\infty} \bar{U}_i$). So the relations $z = 1$ holds either in the yields of both games or of neither game. Hence the yields are isomorphic, and $\mathcal{P}$ is $(\varnothing, \varnothing)$-enforceable.

A similar argument shows that if $\mathcal{P}$ is $(\varnothing, \varnothing)$-enforceable, then $\mathcal{P}$ is finite-enforceable. So the finite and $(\varnothing, \varnothing)$ codes of rules are equivalent.

Let $P = (U, V)$ be any position with finite support, and let $\bar{P} = P^+ \cup P^-$ be the set of consequences of P. If $u \in P^+$, then $u \in \langle A\rangle^F$, for some finite subset A of U. If $z \in P^-$ then $v \in \langle U, z\rangle^F$, for some $v \in V$, and in fact $v \in \langle A, z\rangle^F$, for some finite $A \subseteq U$. Thus, if $g \in \bar{P}$, there is a finite position

$$Q = (A, B) \subseteq P,$$

with $|B| = 1$, such that $g \in \bar{Q}$. The sets

$$L = \{(g, A) \mid A \in \mathcal{P}_{\mathrm{f}}(F), \quad g \in \langle A\rangle^F\},$$
$$M = \{(g, A, v) \mid A \in \mathcal{P}_{\mathrm{f}}(F), \quad g \in F, \quad v \in \langle A, g\rangle^F\}$$

are recursively enumerable. Clearly,

$$P^+ \leqslant_{\mathrm{e}} U \quad \text{via } L$$

and

$$g \in P^- \quad \Leftrightarrow \quad (g, A, v) \in M \ \text{ for some } A \subseteq U \text{ and some } v \in V.$$

Thus, $(P^+, P^-) \leqslant^* (U, V)$.

If $\{w_1(\mathbf{g}), \ldots, w_s(\mathbf{g})\}$ is a set of words in W, let W_1 denote the subset of W which consists of all words on the elements $w_i^{\pm 1}, \ldots, w_s^{\pm 1}$.

At this point, we are in danger of being caught out by our rather sloppy use of the concept of a 'word'. Strictly speaking a word in W is something of the form

$$g_{i_0}^{\varepsilon_0} g_{i_1}^{\varepsilon_1} \cdots g_{i_n}^{\varepsilon_n},$$

where $n \in \mathbb{N}$, $i_j \geqslant 0$, and $\varepsilon_j = \pm 1$, while a word in W_1 is something of the form

$$w_{i_0}^{\varepsilon_0} w_{i_1}^{\varepsilon_1} \cdots w_{i_n}^{\varepsilon_n},$$

where $n \in \mathbb{N}$, $1 \leqslant i_j \leqslant s$, and $\varepsilon_j = \pm 1$. Since each w_i is itself a word in W, there is a natural correspondence between elements of W_1 and elements of W, where, for example, if

$$w_{i_0} = g_1 g_2 \quad \text{and} \quad w_{i_1} = g_1 g_3^{-1},$$

then

$$w_{i_0} w_{i_1}^{-1} \quad \text{corresponds to} \quad g_1 g_2 g_3 g_1^{-1}.$$

We usually blur the difference between a word of W_1, and the word in W to which it natrually corresponds. In this case we need to make a distinction between them; so, for each word

$$u_1(w_1, \ldots, w_s) \in W_1,$$

we will write u_1^* for the word

$$u_1(w_1(\mathbf{g}), \ldots, w_s(\mathbf{g})) \in W$$

to which u_1 naturally corresponds. To highlight the difference, we use the example above. If

$$u_1(w_1, \ldots, w_s) = w_{i_0} w_{i_1}^{-1},$$

then $u_1^* = g_1 g_2 g_3 g_1^{-1}$ while $u_1(g_0, \ldots, g_s) = g_{i_0} g_{i_1}^{-1} \neq u_1^*$

Now let

$$\begin{aligned} X &= \{u \in P^+ \mid \exists u_1 \in W_1,\ u_1^* = u\} = P^+ \cap W_1^*, \\ Z &= \{v \in P^- \mid \exists v_1 \in W_1,\ v_1^* = v\} = P^- \cap W_1^*. \end{aligned}$$

The set $\{u_1^* \mid u_1 \in W_1\}$ is clearly a recursively enumerable subset of W. (In fact it is recursive, because the length of u_1^* is at least the length of u_1; so, if

$$v \notin \{u_1^* \mid \text{length } u_1 \leqslant \text{length } v\},$$

then $v \notin \{u_1^* \mid u_1 \in W_1\}$.) So the sets

$$R = \{(u_1^*, \{u_1^*\}) \mid u_1 \in W_1\}, \qquad S = \{(v_1^*, \varnothing, v_1^*) \mid v_1 \in W_1\}$$

are recursively enumerable, and we have

$$(X, Z) \leqslant^* (P^+, P^-) \quad \text{via } (R, S).$$

We can now prove the following result.

Lemma 7.2 *Let G be a finitely generated group, and let Y be the yield of a game played under the (A, B)-code of rules. If*:

(*i*) (Rel G, Nonrel G) $\leqslant^* (A, B)$, *then* $G \in \mathrm{Sk}\, Y$ *is* (A, B)*-enforceable*;

(*ii*) (Rel G, Nonrel G) $\not\leqslant^* (A, B)$, *then* $G \notin \mathrm{Sk}\, Y$ *is* (A, B)*-enforceable.*

Proof Suppose that (Rel G, Nonrel G) $\leqslant^* (A, B)$. In his first move, the first player can only have used finitely many of the g_i, say $g_0, \ldots, g_m$. Since G is finitely generated, we can pick $g_{m+1}, \ldots, g_n$ to correspond to the generators of G. Let G_1 be the group, generated by $g_{m+1}, \ldots, g_n$, which is isomorphic to G. Then, if the position $P_0 = (U, V)$ is facing us, the position $P_1 = (U \cup \mathrm{Rel}\, G_1, V \cup \mathrm{Nonrel}\, G_1)$ is consistent and (A, B)-legal. We make the move (P_0, P_1), ensuring that $G_1 \leqslant Y$ and hence that $G \in \mathrm{Sk}\, Y$.

We now suppose that (Rel G, Nonrel G) $\not\leqslant^* (A, B)$. Let $G = \langle a_1, a_2, \ldots, a_r \rangle$ and let F be the free group freely generated by $g_0, g_1, g_2, \ldots$. We can effectively enumerate the set of all sequences

$(w_1, \dots, w_r)$ of elements of F. Thus, it is sufficient to show that, with one (A, B)-legal move, we can destroy the possibility that there is an isomorphism $\theta: G \to Y$ carrying a_i to w_i $(1 \leq i \leq r)$. Suppose that we are faced with the position $P = (U, V)$. Let W_1 be the set of all words on $w_1, \dots, w_r$ and their inverses. We can, and we will, think of $\mathrm{Rel}\,(a_1, \dots, a_r)$ as a set of words in W_1. So

$$\mathrm{Rel}\,(a_1, \dots, a_r) \cup \mathrm{Nonrel}\,(a_1, \dots, a_r) = W_1.$$

Since P is an (A, B)-legal position, $(U, V) \leq^* (A, B)$. Let $X = P^+ \cap W_1$ and let $Z = P^- \cap W_1$, then, from the discussions just above the statement of the lemma, we have

$$(X, Z) \leq^* (P^+, P^-) \leq^* (U, V) \leq^* (A, B).$$

As above, for $u_1 \in W_1$ let

$$u_1^* = u_1\big(w_1(\mathbf{g}), \dots, w_r(\mathbf{g})\big) \in W.$$

For $T \subseteq W_1$, let $T^* = \{u_1^* \mid u_i \in T_1\}$. Then W_1 and W_1^* are effectively enumerable sets, and the map $W_1 \to W_1^*$ given by $u_1 \mapsto u_1^*$ is a recursive bijection. So, for all subsets $T \subseteq W_1$, we have

$$T \equiv_1 T^*.$$

If

$$X = \mathrm{Rel}^*\,(a_1, \dots, a_r), \qquad Z = \mathrm{Nonrel}^*\,(a_1, \dots, a_r),$$

then

$$\big(\mathrm{Rel}\,(a_1, \dots, a_r), \mathrm{Nonrel}\,(a_1, \dots, a_r)\big) \equiv_1 (X, Z) \leq^* (A, B),$$

contrary to assumption. So we must have

$$X \neq \mathrm{Rel}^*\,(a_1, \dots, a_r) \quad \text{or} \quad Z \neq \mathrm{Nonrel}^*\,(a_1, \dots, a_r).$$

Suppose first of all that there exists a word $u_1 \in W_1$ such that $u_1^* \notin X \cup Z$. Let

$$Q = \begin{cases} (U \cup \{u_1^*\}, V) & \text{if } u_1 \in \mathrm{Nonrel}\,(a_1, \dots, a_r), \\ (U, V \cup \{u_1^*\}) & \text{if } u_1 \in \mathrm{Rel}\,(a_1, \dots, a_r). \end{cases}$$

In both cases, Q is consistent, because, if $\langle U \cup \{u_1^*\}\rangle^F \cap V \neq \varnothing$, then

$$u_1^* \in P^- \cap W_1^* = Z,$$

by definition, and, if $\langle U\rangle^F \cap (V \cup \{u_1^*\}) = \varnothing$, then

$$u_1^* \in P^+ \cap W_1^* = Z,$$

since $\langle U\rangle^F \cap V = \varnothing$. So Q is an (A, B)-legal position, and (P, Q) is an (A, B)-legal move. We make this move, ensuring that G is not isomorphic to the subgroup of Y generated by $(w_1, \dots, w_r)$.

If $X \cup Z = W_1^*$, then the subgroup of Y generated by $(w_1, \ldots, w_r)$ is already fixed, with

$$\text{Rel}^* (w_1, \ldots, w_r) = X \quad \text{and} \quad \text{Nonrel}^* (w_1, \ldots, w_r) = Z.$$

But since we cannot have

$$\text{Rel}^* (a_1, \ldots, a_r) = X \quad \text{and} \quad \text{Nonrel}^* (a_1, \ldots, a_r) = Z,$$

we must have

$$\text{typ}\,(a_1, \ldots, a_r) \neq \text{typ}\,(w_1, \ldots, w_r).$$

so the proposed isomorphism θ cannot exist anyway, and we play the trivial move. □

We can now see how this theory of games can be used to prove results about existentially closed groups.

Corollary 7.3 (*i*) *If G and H are finitely generated groups with* $\text{Rel}\, H \not\leqslant^* \text{Rel}\, G$, *then there exists a countable existentially closed group M such that $G \in \text{Sk}\, M$ but $H \notin \text{Sk}\, M$.*

(*ii*) *If H is a finitely generated group with unsolvable word problem, then there exists a countable existentially closed group M such that $H \notin \text{Sk}\, M$.*

Proof (i) By definition, if $\text{Rel}\, H \not\leqslant^* \text{Rel}\, G$, then

$$(\text{Rel}\, H, \text{Nonrel}\, H) \not\leqslant^* (\text{Rel}\,(G), \text{Nonrel}\, G).$$

Putting $A = \text{Rel}\, G$ and $B = \text{Nonrel}\, G$ in Lemma 7.2, we see that the properties 'Y is existentially closed', '$H \notin \text{Sk}\, Y$', and '$G \in \text{Sk}\, Y$' are all (A, B)-enforceable. So the required group M is the yield of some game played under the $(\text{Rel}\, G, \text{Nonrel}\, G)$ code of rules.

(ii) If $G = \{1\}$, then $\text{Rel}\, H \leqslant^* \text{Rel}\, G$ if and only if $\text{Rel}\, G$ is recursive, i.e. if and only if H has solvable word problem. So, putting $G = \{1\}$ in (i) forces $\text{Rel}\, H \not\leqslant^* \text{Rel}\, G$, if H has unsolvable word problem, giving the required group, M. □

The question naturally arises, as to which non-trivial non-empty isomorphism-closed classes $\mathscr{X}$ of finitely generated groups form the skeletons of existentially closed groups. We have seen (c.f. Theorem 3.9) that any skeleton of an existentially closed group must satisfy the properties SC, JEP, and AC, and that these properties are also sufficient if $\mathscr{X}$ contains only countably many isomorphism types. Since the skeleton of a countable existentially closed group contains only countably many isomorphism types, this answers the question for countable groups. In general, for any existentially closed group M, we have seen that, if $G \in Sk\, M$ and if H is a finitely generated group such that $\text{Rel}\, H \leqslant^* \text{Rel}\, G$, then $H \in \text{Sk}\, M$. This fact, together with Corollary 7.3(i), suggests that we ask whether closure under $\leqslant^*$ is a necessary and sufficient condition for any non-trivial non-empty isomorphism-closed class $\mathscr{X}$ to be the skeleton of some existentially closed group. However, the answer to this question is 'no'! In fact, Ziegler (1980) has a necessary and sufficient condition for $\mathscr{X}$ to be of the form $\text{Sk}\, M$, which requires closure under $\leqslant^*$, but also involves a condition on the recursively enumerable 'Horn classes' of $\mathbb{N}$. Although the nature of Ziegler's result puts it outside the intended scope of this book, we felt that it was appropriate to mention it here, in order not to mislead the reader into thinking that perhaps closure under $\leqslant^*$ alone might be sufficient.

Theorem 7.4 *There exists an uncountable set*

$$\{[M_\alpha] \mid \alpha \in \Lambda\}$$

of isomorphism classes of countable existentially closed groups such that, if $\alpha, \beta \in \Lambda$ *and* $\alpha \neq \beta$, *then* $(\text{Sk}\, M_\alpha) \cap (\text{Sk}\, M_\beta)$ *consists precisely of the finitely generated groups with solvable word problem.*

Proof Suppose all such sets of classes are countable. Ordering the sets under inclusion, the union of all the sets in a chain is an upper bound for the chain, so that there is a maximum set $X = \{[M_\alpha] \mid \alpha \in \Lambda\}$ (say), where Λ is countable. Each M_α is countable, and so has only countably many finitely generated subgroups. Thus, the set

$$\Pi = \{G \mid G \text{ finitely generated, } G \text{ has unsolvable word problem, } G \subseteq M_\alpha, \text{ for some } \alpha \in \Lambda\}$$

is countable. The properties '$G \notin \text{Sk}\, Y$', for $G \in \Pi$, and 'Y is existentially closed', are $(\varnothing, \varnothing)$-enforceable. This is a total of only countably many properties, all of which are $(\varnothing, \varnothing)$-enforceable. So we can find some existentially closed group Y that is the yield of some game, played under the $(\varnothing, \varnothing)$-rules, whose skeleton contains all of the finitely generated groups with solvable word problem (by Theorem 5.8) and none of the groups in Π. So, by Theorem 5.10, $[Y] \notin X$, contradicting the maximality of X. Hence Λ must be uncountable. □

It is a theorem of Shelah that the set in Theorem 7.4 can be taken to have cardinality $2^{\aleph_0}$. Theorem 7.4 has been included because its proof is relatively simple. To prove Shelah's theorem we need Lemma 7.5, and to state this lemma it is convenient to have the following notation and terminology.

For any position $Q = (U, V)$, let $\Gamma(Q)$ denote the set of all groups K that are generated by $g_0, g_1, g_2, \ldots$ and in which the relations $\{u = 1 \mid u \in U\}$ and the non-relations $\{v \neq 1 \mid v \in V\}$ all hold. So, if $P \subseteq Q$, then $\Gamma(Q) \subseteq \Gamma(P)$.

Let W be the set of all words on the g_i and their inverses. If K is any group generated by $g_0, g_1, g_2, \ldots,$ and if $(w_1, \ldots, w_r) \in W^r$ is any sequence of elements of W, we denote by

$$\langle w_1, \ldots, w_r \rangle_K$$

the subgroup of K generated by the elements of K corresponding to the words w_i $(1 \leq i \leq r)$. We shall think of $\text{Rel}\,(\langle w_1, \ldots, w_r \rangle_K)$ as that set of words $w(w_1, \ldots, w_r)$ such that the relation $w(w_1, \ldots, w_r) = 1$ holds in K.

If $\Delta = ((u_1, \ldots, u_r), (w_1, \ldots, w_r)) \in W^r \times W^r$ is a pair of sequences, we say that Δ is *excluded* by the set $\{Q_1, \ldots, Q_k\}$ of finite positions if, for any $K_i \in \Gamma(Q_i)$ and for any $K_j \in \Gamma(Q_j)$, with $1 \leq i \neq j \leq k$, the existence of an isomorphism

$$\theta : \langle u_1, \ldots, u_r \rangle_{K_i} \to \langle w_1, \ldots, w_r \rangle_{K_j}$$

carrying u_l to w_l $(1 \leq l \leq r)$ implies that $\langle u_1, \ldots, u_r \rangle_{K_i}$ has solvable word problem. In this context, we will say that the isomorphism θ, above, is *unwanted* if it carries u_l to w_l $(1 \leq l \leq r)$ and if $\langle u_1, \ldots, u_r \rangle_{K_i}$ has unsolvable word problem.

Lemma 7.5 *Let $P_1, \ldots, P_k$ be finite positions, and let*

$$\Delta = ((u_1, \ldots, u_r), (w_1, \ldots, w_r))$$

be a pair of sequences in W^r. Then there exist finite positions $Q_i \supseteq P_i$ $(1 \leq i \leq k)$ such that $\{Q_1, \ldots, Q_k\}$ excludes Δ.

Proof We are required to construct Q_i $(1 \leq i \leq k)$ such that, if $K_i \in \Gamma(Q_i)$ and $K_j \in \Gamma(Q_j)$, with $i \neq j$, then there is no unwanted isomorphism

$$\theta : \langle u_1, \ldots, u_r \rangle_{K_i} \to \langle w_1, \ldots, w_r \rangle_{K_j}.$$

We do this by induction on k.

Suppose that $k = 2$, that $P_1 = (U_1, V_1)$, and that $P_2 = (U_2, V_2)$. First we show that we can choose finite positions $T_1 \supseteq P_1$ and $T_2 \supseteq P_2$ such that, if $K_1 \in \Gamma(T_1)$ and $K_2 \in \Gamma(T_2)$, then there is no unwanted isomorphism

$$\theta : \langle u_1, \ldots, u_r \rangle_{K_1} \to \langle w_1, \ldots, w_r \rangle_{K_2}.$$

Let E be the set of words on new variables $x_1, \ldots, x_r$, and their inverses. Then the maps

$$\varphi_1 : E \to W, \qquad \varphi_2 : E \to W,$$

given by

$$x_l^\varepsilon \varphi_1 = u_l^\varepsilon, \quad x_l^\varepsilon \varphi_2 = w_l^\varepsilon \qquad (1 \leq l \leq r,\ \varepsilon = \pm 1),$$

$$(yz)\varphi_1 = y\varphi_1 \cdot z\varphi_1, \quad (yz)\varphi_2 = y\varphi_2 \cdot z\varphi_2, \qquad (y, z \in E),$$

are recursive and injective.

Let

$$N_\alpha = \{z \in E \mid z\varphi_\alpha \in P_\alpha^+\}, \quad M_\alpha = \{z \in E \mid z\varphi_\alpha \in P_\alpha^-\} \qquad (1 \leqslant \alpha \leqslant 2).$$

We consider three cases:

(i) there exists some $y \in (M_2 \cup N_2) \backslash (M_1 \cup N_1)$,

(ii) $M_2 \cup N_2 \subseteq M_1 \cup N_1$, but there exists some $y \notin M_1 \cup N_1$,

(iii) $M_1 \cup N_1 = E$.

In case (i), we take

$$T_1 = \begin{cases} (U_1 \cup \{y\varphi_1\}, V_1) & \text{if } y \in M_2 \\ (U_1, V_1 \cup \{y\varphi_1\}) & \text{if } y \in N_2. \end{cases}$$

If $y\varphi_1 \notin P_1^-$ then $\langle U_1, y\varphi_1 \rangle^F \cap V_1 = \varnothing$, and if $y\varphi_1 \notin P_1^+$ then $\langle U_1 \rangle^F \cap (V_1 \cup \{y\varphi_1\}) = \varnothing$, since (U_1, V_1) is consistent. So, since $y\varphi_1 \notin \bar{P}_1$, the pair T_1 is consistent, and hence is a finite position. We take $T_2 = P_2$. Then, if $K_\alpha \in \Gamma(T_\alpha)$ $(1 \leqslant \alpha \leqslant 2)$, we have, by definition, that the set

$$\{u = 1, \quad v \neq 1 \mid u \in T_\alpha^+, \quad v \in T_\alpha^-\}$$

holds in K_α. So, if $y \in M_2$, the relation $y\varphi_1 = 1$ holds in $\langle w_1, \dots, w_r \rangle_{K_1}$ and $y\varphi_2 \neq 1$ holds in $\langle w_1, \dots, w_r \rangle_{K_2}$. If $y \in N_2$, then $y\varphi_1 \neq 1$ holds in $\langle u_1, \dots, u_r \rangle_{K_1}$ and $y\varphi_2 = 1$ holds in $\langle w_1, \dots, w_r \rangle_{K_2}$. Thus, in either case, there cannot be an isomorphism θ carrying u_l to w_l $(1 \leqslant l \leqslant r)$, for $y\varphi_1\theta \neq y\varphi_2$.

In case (ii), we have $y\varphi_\alpha \notin P_\alpha^+ \cup P_\alpha^-$ $(1 \leqslant \alpha \leqslant 2)$, so we can take

$$T_1 = (U_1 \cup \{y\varphi_1\}, V_1), \qquad T_2 = (U_2, V_2 \cup \{y\varphi_2\}),$$

since these are consistent finite positions. For $K_\alpha \in \Gamma(T_\alpha)$, we have that $y\varphi_1 = 1$ holds in $\langle u_1, \dots, u_r \rangle_{K_1}$ and that $y\varphi_2 \neq 1$ holds in $\langle w_1, \dots, w_r \rangle_{K_2}$. So again there can be no isomorphism $\theta : u_l \mapsto w_l$ $(1 \leqslant l \leqslant r)$.

In case (iii), we can take

$$T_1 = P_1, \qquad T_2 = P_2,$$

because, for all $K_1 \in \Gamma(P_1)$, the group $\langle u_1, \dots, u_r \rangle_{K_1}$ has solvable word problem. To see this, we take $\mathrm{Rel}(\langle u_1, \dots, u_r \rangle_{K_1})$ and $\mathrm{Rel}(\langle u_1, \dots, u_r \rangle_{K_2})$ both to be subsets of E. So, for $z \in E$ and for $\alpha = 1, 2$,

$$z\varphi_\alpha \in T_\alpha^+ \Rightarrow z\varphi_\alpha = 1 \text{ in } K_\alpha,$$

$$z\varphi_\alpha \in T_\alpha^- \Rightarrow z\varphi_\alpha \neq 1 \text{ in } K_\alpha.$$

Then, since

$$N_1 \subseteq \mathrm{Rel}(\langle u_1, \dots, u_r \rangle_{K_1}), \qquad M_1 \subseteq \mathrm{Nonrel}(\langle u_1, \dots, u_k \rangle_{K_1}),$$

and $M_1 \cup N_1 = E$, we must have $N_1 = \mathrm{Rel}(\langle u_1, \dots, u_r \rangle_{K_1})$. So we have that

$$z \in \mathrm{Rel}(\langle u_1, \dots, u_r \rangle_{K_1}) \quad \text{if and only if} \quad z\varphi_1 \in P_1^+.$$

Since P_1 is finite, P_1^+ and P_1^- are recursively enumerable. So we can list P_1^+ and P_1^- until $z\varphi_1$ appears. If we find that $z\varphi_1 \in P_1^+$, then

$$z \in \mathrm{Rel}\,(\langle u_1, \dots, u_r \rangle_{K_1}),$$

but, if $z\varphi_1 \in P_1^-$, then $z \in M_1 = \mathrm{Nonrel}\,(\langle u_1, \dots, u_r \rangle_{K_1})$, and $z \notin \mathrm{Rel}\,(\langle u_1, \dots, u_r \rangle_{K_1})$. So $\mathrm{Rel}\,(\langle u_1, \dots, u_r \rangle_{K_1})$ is recursive, and there can be no unwanted isomorphism

$$\theta : \langle u_1, \dots, u_r \rangle_{K_1} \to \langle w_1, \dots, w_r \rangle_{K_2}.$$

We now repeat this process with the positions T_2 and T_1, and find positions

$$Q_2 \supseteq T_2, \qquad Q_1 \supseteq T_1$$

such that, if $K_2 \in \Gamma(Q_2)$ and $K_1 \in \Gamma(Q_1)$, then there is no unwanted isomorphism

$$\theta : \langle u_1, \dots, u_r \rangle_{K_2} \to \langle w_1, \dots, w_r \rangle_{K_1}.$$

We note that, if $K_\alpha \in \Gamma(Q_\alpha)$, then $K_\alpha \in \Gamma(T_\alpha)$ $(1 \leqslant \alpha \leqslant 2)$. So there can be no unwanted isomorphism

$$\theta : \langle u_1, \dots, u_r \rangle_{K_1} \to \langle w_1, \dots, w_r \rangle_{K_2}.$$

Thus $\{Q_1, Q_2\}$ excludes Δ, as required.

If $k \geqslant 3$, then, by induction, we can choose finite positions $R_j \supseteq P_j$ $(1 \leqslant j \leqslant k-1)$ such that $\{R_1, \dots, R_{k-1}\}$ excludes Δ. Also, by induction, we can choose $S_k \supseteq P_k$ and $S_j \supseteq R_j$ $(2 \leqslant j \leqslant k-1)$ such that $\{S_2, \dots, S_k\}$ excludes Δ. Finally, we can choose $Q_1 \supseteq R_1$ and $Q_j \supseteq S_j$ $(3 \leqslant j \leqslant k)$ such that

$$\{Q_1, Q_3, \dots, Q_k\}$$

excludes Δ.

Now let $Q_2 = S_2$, and suppose that

$$K_i \in \Gamma(Q_i) \quad \text{and} \quad K_j \in \Gamma(Q_j) \qquad (1 \leqslant i \neq j \leqslant k).$$

If $2 \leqslant i,j \leqslant k$, then

$$K_i \in \Gamma(S_i) \quad \text{and} \quad K_j \in \Gamma(S_j),$$

so there is no unwanted isomorphism

$$\theta : \langle u_1, \dots, u_r \rangle_{K_i} \to \langle w_1, \dots, w_r \rangle_{K_j}.$$

If $1 \leqslant i,j \leqslant k-1$, then $K_i \in \Gamma(R_i)$ and $K_j \in \Gamma(R_j)$, so again there are no unwanted isomorphisms. The only other possibility is that $\{i, j\} = \{1, k\}$; in this case, there is no unwanted isomorphism, since $\{Q_1, Q_3, \dots, Q_k\}$ excludes Δ. Thus, $\{Q_1, Q_2, \dots, Q_k\}$ excludes Δ and is the required set. The result thus follows by induction. □

Theorem 7.6 *There exist $2^{\aleph_0}$ pairwise non-isomorphic countable existentially closed groups $\{M_\alpha \mid \alpha \in 2^{\aleph_0}\}$ such that, if $\alpha \neq \beta$, then $(\mathrm{Sk}\, M_\alpha) \cap (\mathrm{Sk}\, M_\beta)$ consists precisely of the finitely generated groups with soluble word problem.*

Proof Since there are, up to isomorphism, only $2^{\aleph_0}$ countable groups, our task is to find $2^{\aleph_0}$ countable existentially closed groups with the required property. Each of these will contain at least one finitely generated group with unsolvable word problem, by Theorem 5.10. All the groups will thus have different skeletons, and hence no two will be isomorphic, giving the result.

We consider the binary tree T (see Fig. 7.1).

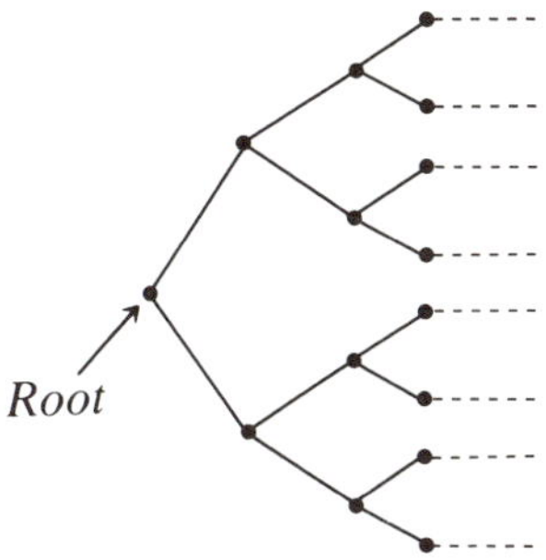

Figure 7.1

A branch of T is an infinite collection of points, starting at the root, which are joined in successive pairs (no point appearing twice in any one branch). Since each branch can be represented uniquely by an infinite sequence of 0's and 1's, there are $2^{\aleph_0}$ branches. The root is, by definition, the first point on every branch, and points will be called even or odd depending on whether they are in an even or odd position along a branch. (This naming is clearly independent of which branch the point is considered to be lying on.)

Our aim is to attach a finite position P_v to every point v of T in such a way that, if we read the positions consecutively along branches, we get games that have been played under the finite rules, with yields that are all non-isomorphic and of the type we require.

We can turn a second-player strategy into a first-player strategy by assuming that the first position was $(\varnothing, \varnothing)$. Thus, using a first-player strategy, we choose positions at all the odd points of each branch, as we come to them, to ensure that the yield is existentially closed. We now describe how to choose the positions at the even points to ensure that our yields are subgroup-skew (i.e. that the skeletons intersect in exactly the finitely generated groups with solvable word problem).

We enumerate all the triples $(v, (u_1, \dots, u_r), (w_1, \dots, w_r))$, where v is a vertex of T, $r \in \mathbb{N}$, and u_j and w_j are words in W $(1 \leq j \leq r)$. The second player deals with the ith triple on his ith move (the $2i$th move of the game). We can choose our enumeration such that, if v belongs to the ith triple, then v is the jth vertex on a branch, for some $j < i$. There are $k = 2^{2i-2}$ vertices of T which are the $(2i-1)$th vertices of the branches. To these vertices are assigned positions $P_1, P_2, \dots, P_k$. Let

$$\Delta = ((u_1, \dots, u_r), (w_1, \dots, w_r)).$$

Then, since $P_1, P_1, P_2, P_2, \dots, P_k, P_k$ are finite positions, by Lemma 7.5, we can find finite positions Q_i $(1 \leq i \leq 2k)$ such that $P_j \subseteq Q_{2j-1}$ and $P_j \subseteq Q_{2j}$ $(1 \leq j \leq k)$, and such that $\{Q_1, \dots, Q_{2k}\}$ excludes Δ. We assign the positions Q_i to the $2k$ vertices of T that lie at a $2i$th point along some branch (see Fig. 7.2).

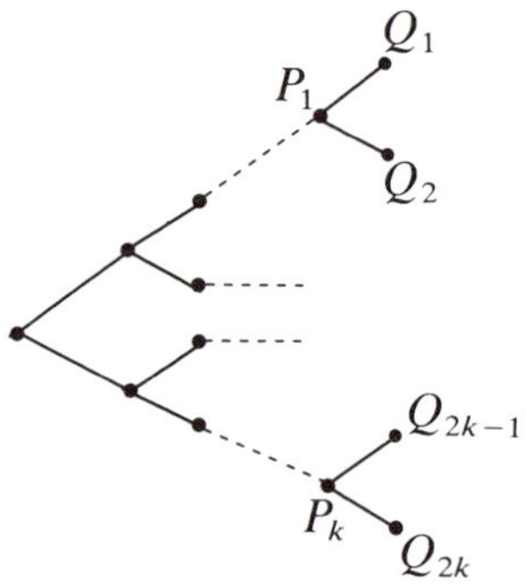

Figure 7.2

In this way we get $2^{\aleph_0}$ finite games, the yield of each of which is an existentially closed group. Thus we get $2^{\aleph_0}$ existentially closed groups, corresponding to the branches of T, all of whose skeletons contain every finitely generated group with solvable word problem, by Theorem 5.8, and, by Theorem 5.10, at least one finitely generated group with unsolvable word problem. Suppose Y_p and Y_q are two of these groups, corresponding to distinct branches B_p and B_q.

Let $G_p \subseteq Y_p$ and $G_q \subseteq Y_q$ be isomorphic groups, let $G_p = \langle u_1, \dots, u_r \rangle_{Y_p}$, let $G_q = \langle w_1, \dots, w_r \rangle_{Y_q}$, and let $\theta: G_p \to G_q$ be an isomorphism carrying u_i to w_i $(1 \leq i \leq r)$. Since B_p and B_q are distinct branches, they must divide at some vertex v. Suppose that

$$(v, (u_1, \dots, u_r), (w_1, \dots, w_r))$$

is the ith triple. Let v_p and v_q be the $2i$th vertices on B_p and B_q, respectively. Then, by choice of enumeration, v lies before v_p and v_q. Thus $v_p \neq v_q$. If Q_p and Q_q are the positions associated with v_p and v_q, then $p \neq q$ and, since Q_p is a position in a game which yields Y_p, we have

$Y_p \in \Gamma(Q_p)$, and similarly $Y_q \in \Gamma(Q_q)$. But $\{Q_1, \ldots, Q_{2k}\}$ excludes

$$\Delta = ((u_1, \ldots, u_r), (w_1, \ldots, w_r)),$$

by construction, so there is no unwanted isomorphism

$$\theta: \langle u_1, \ldots, u_r \rangle_{Y_p} \to \langle w_1, \ldots, w_r \rangle_{Y_q}.$$

Since $G_p = \langle u_1, \ldots, u_r \rangle_{Y_p}$ and θ is an isomorphism, G_p must have solvable word problem. Thus $\mathrm{Sk}\, Y_p \cap \mathrm{Sk}\, Y_q$ contains only groups with solvable word problem, as required. □

Note Theorem 7.6 follows directly from Theorem 7.4 if we assume the continuum hypothesis. Our proofs, of course, do not assume it.

Problem We would like to prove a modification of Theorem 7.6 which would say that, given any finitely generated group G, there exist $2^{\aleph_0}$ isomorphism classes $[M_\alpha]$ of countable existentially closed groups such that, if $[M_\alpha] \neq [M_\beta]$, then

$$(\mathrm{Sk}\, M_\alpha) \cap (\mathrm{Sk}\, M_\beta) = \{H \mid \mathrm{Rel}\, H \leqslant^* \mathrm{Rel}\, G\}.$$

It is not hard to see that we can modify Lemma 7.5 and Theorem 7.6 to consider games played under the (Rel G, Nonrel G) code of rules. We say that an isomorphism

$$\theta: \langle h_1, \ldots, h_r \rangle \to \langle k_1, \ldots, k_r \rangle$$

is unwanted if $\mathrm{Rel}\,(h_1, \ldots, h_r) \not\leqslant^* \mathrm{Rel}\, G$. Then we prove that, given positions $P_1, \ldots, P_k$, we find (Rel G, Nonrel G)-legal positions $Q_i \supseteq P_i$ so that

$$\{Q_1, \ldots, Q_k\} \quad \text{excludes} \quad ((h_1, \ldots, h_r), (k_1, \ldots, k_r)).$$

We then construct positions on a binary tree, as in Theorem 7.6, but we assign the position (Rel G, Nonrel G) to the root, so that G belongs to the yields of all the branches. If M_α denotes the yield of B_α, and if $\alpha \neq \beta$, then by Theorem 5.7 and by construction,

$$(\mathrm{Sk}\, M_\alpha) \cap (\mathrm{Sk}\, M_\beta) = \{H \mid \mathrm{Rel}\, H \leqslant^* \mathrm{Rel}\, G\}.$$

The problem is that we do not know that $\mathrm{Sk}\, M_\alpha$ and $\mathrm{Sk}\, M_\beta$ are distinct, and hence that $M_\alpha \not\equiv M_\beta$, because we do not have an analogue of Theorem 5.10. If we could prove that the property

'$H \in \mathrm{Sk}\, Y$ for some H such that $\mathrm{Rel}\, H \not\leqslant^* \mathrm{Rel}\, G$'

were (Rel G, Nonrel G)-enforceable, then we could force $\mathrm{Sk}\, M_\alpha \neq \mathrm{Sk}\, M_\beta$ and get the result. However, we have been unable to do this.

8
Free products

This chapter is devoted to proving the following theorem.

Theorem 8.1 *There exists a countable existentially closed group M and a finitely generated group $G \in \mathrm{Sk}\, M$ such that*

$$G * H \notin \mathrm{Sk}\, M$$

for any non-trivial group H. Moreover, we can choose such a G to have a recursively enumerable relation set.

For any effectively enumerable set W, and for $X \subseteq W$, let

$$\mathrm{s}(X; W) = \{A \in \mathcal{P}_{\mathrm{f}}(W) \mid A \cap X \neq \varnothing\}.$$

In the case $W = \mathbb{N}$, we may just write $\mathrm{s}(X)$ for $\mathrm{s}(X; \mathbb{N})$.

Lemma 8.2 *If $G = \langle a_1, \ldots, a_r \rangle$ is a finitely generated group, and if $\langle t \rangle$ is an infinite cyclic group with $G \cap \langle t \rangle = \{1\}$, then*

$$\mathrm{s}(\mathrm{Rel}\, G) \leqslant_1 \mathrm{Rel}\,(G * \langle t \rangle).$$

Proof We consider the set $W = \mathrm{W}(x_1, \ldots, x_r)$ of all words on $x_1, \ldots, x_r$ and their inverses, and we suppose that this set is ordered, somehow, so that if 1 is the empty word then $1 < w$, for all $w \in W \backslash \{1\}$; and if $w_1 \neq w_2$ then either $w_1 < w_2$ or $w_2 < w_1$, (but not both). Let y be a new element not in $\mathrm{W}(x_1, \ldots, x_r)$. Define a function

$$f : \mathcal{P}_{\mathrm{f}}(W) \to \mathrm{W}(x_1, \ldots, x_r, y)$$

inductively as follows:

$$f(\{w_1\}) = f(w_1) = w_1,$$

$$f(\{w_1, \ldots, w_k\}) = f(w_1, \ldots, w_k) = [w_k, y^{-1} f(w_1, \ldots, w_{k-1}) y] \quad (k \geqslant 2),$$

where $w_k > w_i$ $(1 \leqslant i \leqslant k-1)$. (We need to choose w_k in some fixed position in the ordering of $\{w_1, \ldots, w_k\}$ in order to ensure that f is well defined.) Here, $[x, y]$ is the commutator $x^{-1}y^{-1}xy$ and, since we are considering sets of words, w^{-1} denotes the formal inverse of w. Clearly, f is recursive and, since we are considering sets of words, and not free groups, we see by induction on the number of occurrences of y in $f(w_1, \dots, w_k)$ that f is injective. We take $\mathrm{Rel}\,(a_1, \dots, a_r)$ to be in W and $\mathrm{Rel}\,(a_1, \dots, a_r, t)$ to be in $\mathrm{W}(x_1, \dots, x_r, y)$, and we show that

$$\mathrm{s}\,(\mathrm{Rel}\,(a_1, \dots, a_r); W) \leqslant_1 \mathrm{Rel}\,(a_1, \dots, a_r, t) \qquad \text{via } f.$$

Then, since $\mathrm{Rel}\,(G * \langle t \rangle) \equiv_1 \mathrm{Rel}\,(a_1, \dots, a_r, t)$ and since

$$\mathrm{s}\,(\mathrm{Rel}\,(a_1, \dots, a_r); W) \equiv_1 \mathrm{s}\,(\mathrm{Rel}\,(G); \mathbb{N}),$$

we will have the result.

If $\{w_1, \dots, w_k\} \in \mathrm{s}\,(\mathrm{Rel}\,(a_1, \dots, a_r); W)$, then w_i belongs to $\mathrm{Rel}\,(a_1, \dots, a_r)$, for some $i \in \{1, \dots, k\}$. Suppose that $w_1 < w_2 < \cdots < w_k$, and let $u_j = f(w_1, \dots, w_{j-1})$ $(2 \leqslant j \leqslant k+1)$, so that

$$f(w_1, \dots, w_j) = w_j^{-1} y^{-1} u_j^{-1} y w_j y^{-1} u_j y = u_{j+1}.$$

So, since $\mathrm{Rel}\,(a_1, \dots, a_r) \subseteq \mathrm{Rel}\,(a_1, \dots, a_r, t)$ and since $w_i \in \mathrm{Rel}\,(a_1, \dots, a_r)$, it follows that

$$u_{i+1} \in \mathrm{Rel}\,(a_1, \dots, a_r, t).$$

Then

$$u_{i+2}, \dots, u_k \in \mathrm{Rel}\,(a_1, \dots, a_r, t),$$

and so

$$f(w_1, \dots, w_k) \in \mathrm{Rel}\,(a_1, \dots, a_r, t).$$

Conversely, we want to show that if

$$f(w_1, \dots, w_k) \in \mathrm{Rel}\,(a_1, \dots, a_r, t)$$

then

$$\{w_1, \dots, w_k\} \in \mathrm{s}\,\big(\mathrm{Rel}\,(a_1, \dots, a_r); W\big)$$

If $k = 1$, then $f(w_1) = w_1$, so $f(w_1) \in \mathrm{Rel}(a_1, \dots, a_r, t)$ implies that

$$w_1 \in \mathrm{Rel}\,(a_1, \dots, a_r),$$

and thus that $\{w_1\} \in \mathrm{s}\,\big(\mathrm{Rel}\,(a_1, \dots, a_r); W\big)$. If $k \geqslant 2$, then

$$f(w_1, \dots, w_k) = w_{l_1}^{\varepsilon_1} y^{n_1} \cdots w_{l_m}^{\varepsilon_m} y^{n_m},$$

where $l_i \in \{1, \dots, k\}$, $\varepsilon_i = \pm 1$, $n_i \in \mathbb{Z} \backslash \{0\}$, and $n_m > 0$. By the Normal-Form Theorem for free products, if

$$f(w_1, \dots, w_k) \in \mathrm{Rel}\,(a_1, \dots, a_r, t)$$

then $w_{l_i} \in \mathrm{Rel}\,(a_1, \dots, a_r)$, for some i. So

$$\{w_1, \dots, w_k\} \in \mathrm{s}\,\big(\mathrm{Rel}\,(a_1, \dots, a_r); W\big),$$

and f is the required injective recursive function. □

Corollary 8.3 *To prove the weak form (first sentence) of Theorem* 8.1 *it is sufficient to prove that there exists some* $X \subseteq \mathbb{N}$ *such that* $\mathrm{s}(X) \not\leq^* X$. *Moreover, the full theorem will follow if this* X *can be chosen to be recursively enumerable.*

Proof Suppose that we have a set X such that $\mathrm{s}(X) \not\leq^* X$. We can construct the group G_X, such that $\mathrm{Rel}\,\mathrm{G}_X \equiv^* X$, (see Chapter 4). From Theorem 4.23 we see that there is some injective recursive function $g : \mathbb{N} \to \mathbb{N}$, such that $n \in X$ if and only if $g(n) \in \mathrm{Rel}\, G_X$. We define $f : \mathscr{P}_{\mathrm{f}}(\mathbb{N}) \to \mathscr{P}_{\mathrm{f}}(\mathbb{N})$ by the rule

$$f(\{n_1, \ldots, n_m\}) = \{g(n_1), \ldots, g(n_m)\}.$$

Then f is also injective and recursive, and $\{n_1, \ldots, n_m\} \in \mathrm{s}(X)$ if and only if $n_i \in X$, for some $i \in \{1, \ldots, m\}$, which is if and only if $g(n_i) \in \mathrm{Rel}\,\mathrm{G}_X$, and hence if and only if

$$\{g(n_1), \ldots, g(n_m)\} \in \mathrm{s}(\mathrm{Rel}\,\mathrm{G}_X).$$

So $\mathrm{s}(X) \leq_1 \mathrm{s}(\mathrm{Rel}\,\mathrm{G}_X)$. We show that G_X is the finitely generated group required for Theorem 8.1.

First we will consider the free product $\mathrm{G}_X * \langle t \rangle$ of G_X with an infinite cyclic group. If

$$\mathrm{Rel}\,(\mathrm{G}_X * \langle t \rangle) \leq^* \mathrm{Rel}\,\mathrm{G}_X$$

then, using Lemma 8.2, we get

$$\mathrm{s}(X) \leq^* \mathrm{s}(\mathrm{Rel}\,\mathrm{G}_X) \leq^* \mathrm{Rel}\,(\mathrm{G}_X * \langle t \rangle) \leq^* \mathrm{Rel}\,\mathrm{G}_X \equiv^* X,$$

which is contrary to assumption. So $\mathrm{Rel}\,(\mathrm{G}_X * \langle t \rangle) \not\leq^* \mathrm{Rel}\,\mathrm{G}_X$. Thus, by Corollary 7.3(i), there exists a countable existentially closed group M, such that $\mathrm{G}_X \in \mathrm{Sk}\, M$ but $\mathrm{G}_X * \langle t \rangle \notin \mathrm{Sk}\, M$.

Finally, if H is any non-trivial group, then $G_X * H$ contains a non-trivial conjugate of G_X, and G_X contains an element of infinite order. So

$$G_X * \langle t \rangle \subseteq G_X * G_X \subseteq G_X * H,$$

and $G_X * H \notin \mathrm{Sk}\, M$.

Moreover, if X is recursively enumerable, then G_X has a recursively enumerable relation set. Thus we would have Theorem 8.1. □

In the next lemma we deal with the relatively easy case when X is not required to be recursively enumerable. We prove the harder case in the lemma after next.

Lemma 8.4 *There exists some* $X \subseteq \mathbb{N}$ *such that* $\mathrm{s}(X) \not\leq^* X$.

Proof Let V be any subset of the set of triples,

$$\mathscr{P}_f(\mathbb{N}) \times \mathscr{P}_f(\mathbb{N}) \times (\mathbb{N} \cup \{\infty\}).$$

Since the set

$$u = \{(A, \{n\}) \mid A \in \mathscr{P}_f(\mathbb{N}), \quad n \in A\}$$

is recursively enumerable, and since $A \in \mathrm{s}(X)$ if and only if $(A, \{n\}) \in U$ for some $n \in X$, we have that $\mathrm{s}(X) \leqslant_e X$, for all $X \subseteq \mathbb{N}$. Thus,

$$\mathrm{s}(X) \leqslant^* X$$

if and only if, for some recursively enumerable set V of triples in $\mathscr{P}_f(\mathbb{N}) \times \mathscr{P}_f(\mathbb{N}) \times (\mathbb{N} \cup \{\infty\})$, we have

$$A \notin \mathrm{s}(X) \Leftrightarrow (A, B, n) \in V,$$

for some $B \subseteq X$ and for some $n \notin X$. In this case, we shall say that V *works for* X. Our aim, then, is to construct a set X such that no V works for X. In fact, we prove the lemma by showing that, if A and B are disjoint finite subsets of $\mathbb{N}$, and if V is any set of triples, then we can find disjoint finite subsets $A^*, B^* \subseteq \mathbb{N}$, with $A \subseteq A^*$ and $B \subseteq B^*$, such that V does not work for any set X with $A^* \subseteq X$ and $B^* \cap X = \varnothing$. From this we will deduce the existence of some X of the required form.

Given A, B, and V, as above, choose $c, d \in \mathbb{N} \backslash (A \cup B)$ such that $c \neq d$. We look at the triples in V of the form $(\{c, d\}, C, n)$.

Case 1 For one of these triples, $C \cap B = \varnothing$, $n \notin C \cup A$, and, we may suppose, $c \neq n$:
Put $A^* = A \cup C \cup \{c\}$ and put $B^* = B \cup \{n\}$. Then $A^* \cap B^* = \varnothing$ and, if $A^* \subseteq X$ and $B^* \cap X = \varnothing$, then $C \subseteq X$ and $n \notin X$ but $\{c, d\} \in \mathrm{s}(X)$, since $c \in X$. So V cannot work for X.

Case 2 For every such triple either $C \cap B \neq \varnothing$ or $n \in C \cup A$:
In this case, put $A^* = A$ and $B^* = B \cup \{c, d\}$. If $A^* \subseteq X$ and $B^* \cap X = \varnothing$, then $\{c, d\} \subseteq \mathbb{N} \backslash X$. So if V works for X, then $(\{c, d\}, C, n) \in V$, for some $C \subseteq X$ and for some $n \notin X$. But then $C \cap B = \varnothing$ and $n \notin C \cup A \subseteq X$, contrary to assumption. So V cannot work for X.

There are only countably many recursively enumerable subsets of $\mathscr{P}_f(\mathbb{N}) \times \mathscr{P}_f(\mathbb{N}) \times (\mathbb{N} \cup \{\infty\})$, so there are only countably many sets $V_0, V_1, V_2, \ldots$ of triples that might work for any set X. We choose A_0^* and B_0^* such that V_0 does not work for any X with $A_0^* \subseteq X$ and $B_0^* \cap X = \varnothing$. Given A_{i-1}^* and B_{i-1}^*, we choose $A_i^* \supseteq A_{i-1}^*$ and $B_i^* \supseteq B_{i-1}^*$ such that V_i does not work for any X with $A_i^* \subseteq X$ and $B_i^* \cap X = \varnothing$. We put $X = \bigcup_{i \in m} A_i^*$. Then, for all j, we have $A_j^* \subseteq X$ and $B_j^* \cap X = \varnothing$, so none of the V_i works for X, and $\mathrm{s}(X) \not\leqslant^* X$. □

Lemma 8.5 *There exists some recursively enumerable set $X \subseteq \mathbb{N}$ such that* $s(X) \not\leq^* X$.

Proof Using the same terminology as in Lemma 8.4, we are required to construct some recursively enumerable set X so that no recursively enumerable subset of $\mathcal{P}_f(\mathbb{N}) \times \mathcal{P}_f(\mathbb{N}) \times (\mathbb{N} \cup \{\infty\})$ works for X.

We start with the effectively enumerated set $\{V_i \mid i \in \mathbb{N}\}$ containing, with repetitions, precisely those subsets of $\mathcal{P}_f(\mathbb{N}) \times \mathcal{P}_f(\mathbb{N}) \times (\mathbb{N} \cup \{\infty\})$ that are recursively enumerable. Thus, the set

$$Y = \{(A, B, m, i) \mid (A, B, m) \in V_i, \quad i \in \mathbb{N}\}$$

is recursively enumerable. We shall construct X stage by stage.

Since Y is recursively enumerable, there is a machine M that constructs the elements of Y. Set M going, and let Y_n denote the (finite) set of all elements of Y produced by M up to the beginning of the nth stage of the construction of X. Let X_n denote the part of X that has been constructed by the beginning of the nth stage of the construction (X_n will also be finite). We take $X_1 = \emptyset$, and, at the end of the construction, X will be $\bigcup_{n \in \mathbb{N}} X_n$.

At stage n of the construction, a quadruple (A, B, m, i) will be called a *potential witness* (PW) against V_i, if

(i) $(A, B, m, i) \in Y_n$,
(ii) $m \notin X_n$,
(iii) $B \subseteq X_n$ and $A \cap X_n \neq \emptyset$.

Thus, if m is not later assigned to X, the fact that $(A, B, m) \in V_i$ will ensure that V_i does not work for X. From time to time during the construction, we shall elect a PW to be a preferred potential witness (PPW). Let P_n denote the set of PPW's at the beginning of the nth stage of the construction. At any time, a set V_i may have many PWs against it, but it will always have at most one PPW against it.

We take a recursive function f that takes every value in $\mathbb{N}$ infinitely many times. At the nth stage of our construction of X, we act to ensure that V_j does not work for X, where $f(n) = j$.

At the beginning of the nth stage of the construction we have:

A finite list Y_n of quadruples (A, B, m, i).
A finite list X_n of elements already assigned to X.
A finite list P_n of PPWs against certain V_i.

First check P_n. If there exists a PPW, in P_n, against V_j, then put $X_{n+1} = X_n$ and $P_{n+1} = P_n$, and go to stage $n + 1$ of the construction.

If there is no PPW against V_j, check Y_n to see if it contains a PW against V_j. If so, we take the first PW, say (A, B, m, j), against V_j, that we come across in Y_n, and elect it to be a PPW. Then we put $X_{n+1} = X_n$ and $P_{n+1} = P_n \cup \{(A, B, m, j)\}$, and go to stage $n + 1$ of the construction.

If Y_n contains no PWs against V_j, then we check through Y_n again, looking for quadruples (A, B, m, j) such that $m \notin X_n$ and such that, for some $k \neq m$,

(a) $k > 2^{j+1}$,

(b) there is no PPW (C, D, k, i) in P_n, for any $i < j$,

(c) $A \cap (X_n \cup \{k\}) \neq \varnothing$,

(d) $B \subseteq (X_n \cup \{k\})$.

(Thus, (A, B, m, j) can be made into a PW against V_j by adding k to X.) If such quadruples exist, we take the first that we come across in Y_n, and the smallest k with the required properties, and we elect (A, B, m, j) to be a PPW. We put $X_{n+1} = X_n \cup \{k\}$, so that $k \in X$. But then quadruples of the form (C, D, k, i) $(i \in \mathbb{N})$ are no longer PWs, so we delete all the PPWs of this form from P_n, and we put

$$P_{n+1} = \big(P_n \backslash \{(C, D, k, i) \mid C, D \in \mathcal{P}_f(\mathbb{N}), \quad i \in \mathbb{N}\}\big) \cup \{(A, B, m, j)\}.$$

We then go to stage $n + 1$ of the construction.

If none of these cases occurs, we just put $X_{n+1} = X_n$ and $P_{n+1} = P_n$, and we go to stage $n + 1$.

In the second and third cases, above, we say that we have *taken action* against V_j; in the first and fourth cases we have taken no action. (So we have taken action against $V_{f(n)}$ at the nth stage of the construction if and only if $P_{n+1} \neq P_n$.)

We now show that the set $X = \bigcup_{n \in \mathbb{N}} X_n$, constructed in this way, is the required set.

Since we have described an effective construction procedure for X, it follows, by Church's Thesis, that X is recursively enumerable. To prove that no V_i works for X, we need the following two remarks.

Whenever we take action against V_j, we create a PPW against V_j. So if, at some later stage, we are in a position where we need to take action against V_j again, we must have destroyed this PPW at some interim stage. If a PPW (A, B, m, j) has been destroyed at stage n, then we must have $j > f(n)$, because, by (b), we only delete (A, B, m, j) if there is no PPW of the form (A, B, m, i) $\big(i < f(n)\big)$. $\big($If $j = f(n)$, then we take no action and (A, B, m, j) is not destroyed.$\big)$ So, in between two stages at which we take action against V_j, we must take action against some V_i, with $i < j$. Thus, if n_i is the number of times we take action against V_i while constructing X, we have that

$$n_j \leqslant 1 + \sum_{i<j} n_i \quad (j \in \mathbb{N}),$$

which, by induction, gives $n_j \leqslant 2^j$. (i.e. we take action against each V_j at most 2^j times, throughout the whole construction of X.)

A number $k \leqslant 2^{j+1}$ is only added to X when action is taken against some V_i, with $i<j$. Thus, at most $\sum_{i<j} n_i$ such k can belong to X, and so

$$|\mathbb{N}\backslash X| \geqslant 2^{j+1} - \sum_{i<j} n_i \geqslant 2^j,$$

for all $j \in \mathbb{N}$. In other words, the complement of X is infinite.

We now show that no V_j works for X.

Given j, the construction has some stage n after which no further action is taken against any V_i such that $i \leqslant j$. We consider two cases.

Case 1 At this stage, there is a PPW (A, B, m, j) against V_j in P_n: In this case, $A \cap X_n \neq \varnothing$, $B \subseteq X_n$, and $m \notin X_n$. So, $A \cap X \neq \varnothing$, $B \subseteq X$, and $m \notin X$, if this PPW is not destroyed at any later stage in the construction.

To destroy this PPW, we must take action against some V_l with $m > 2^{l+1}$, and there must be no PPW in P_n of the form (C, D, m, i), with $i<l$. Since we assumed that we would take no further action against any V_i with $i \leqslant j$, we must get, if we take action against V_l, that $l>j$. But then (A, B, m, j) is a PPW in P_n with $j<l$, so the very existence of (A, B, m, j) prevents us from taking action that would destroy it. Thus, (A, B, m, j) is not destroyed at any later stage in the construction, and $m \notin X$. So $(A, B, m) \in V_j$ with $B \subseteq X$, $m \notin X$, and $A \notin s(X)$, and hence V_j cannot work for X.

Case 2 At this stage, there is no PPW against V_j: We show that this case cannot happen.

Since $\mathbb{N}\backslash X$ is infinite, we can choose

$$A = \{a_0, a_1, \ldots, a_{j+1}\} \subseteq \mathbb{N}\backslash X$$

such that $a_i > 2^{j+1}$, for all $i \in \{0, \ldots, j+1\}$. If V_j works for X, then there exist $B \subseteq X$ and $m \notin X$ such that $(A, B, m) \in V_j$. We may suppose that, if $m \in A$, then $m = a_{j+1}$, so $m \neq a_i$ $(0 \leqslant i \leqslant j)$.

There exists some integer p such that, at the pth stage of the construction, $B \subseteq X_p$ and $(A, B, m, j) \in Y_p$. Of course, we can take $p \geqslant n$, so that at the pth stage of the construction, all the action against all the V_i, with $i \leqslant j$, has already been taken. By choice of f, there exists some $q \geqslant p$ such that $f(q)=j$. We consider the smallest such q. Since P_n contains no PPW against V_j, and since, by choice of n, we take no action against V_j at any stage after the nth, then P_q contains no PPW against V_j. We show that we are forced to take action against V_j at the qth stage, and hence we get a contradiction.

Since there is at most one PPW in P_q against each V_i, there are at most j PPWs in P_q that are PPWs against some V_i with $i<j$. But

$$|\{a_0, \ldots, a_j\}| = j+1,$$

so we may suppose that there is no PPW of the form (C, D, a_j, i) in P_q, for any $i < j$. Now,

$$a_j \neq m, \qquad a_j > 2^{j+1}, \qquad A \cap (X_q \cup \{a_j\}) \neq \varnothing;$$

so, according to the rules of construction, we should take action against V_j at this stage. This is contrary to the choice of n, since $q \geqslant n$, and so case 2 cannot arise.

This proves that we have constructed a recursively enumerable set X with $\mathrm{s}(X) \leqslant^* X$, as required. □

9
First-order theory of existentially closed groups

This chapter is divided into five sections. We begin by defining a first-order language L, of group theory, and then explain how to assign truth values to statements within the language. We define the first-order theory of a group, and show that, given any existentially closed group M, there exists a group Y which is not existentially closed but which has the same first-order theory as M. In the third section we define the notion of stable truth, and use this to describe a whole class of sentences of L that are true in every existentially closed group. We then give contrasting examples of sentences that are true in some existentially closed groups but not in others. In the fourth section, we prove that if G is a finitely generated group such that Rel G is arithmetic, then there is a first-order sentence f that is true in an existentially closed group M if and only if $G \in \text{Sk}\, M$. In the last section we return to the consideration of those existentially closed groups that can be constructed as the yields of games.

9.1 INTRODUCTION

We consider the first-order language L, which consists of variables $x_0, x_1, x_2, \ldots$; logical symbols &, *or*, $\neg$, $\forall$, $\exists$; a binary relation symbol, $=$; binary and unary function symbols $\cdot$, $^{-1}$; a constant symbol 1; and parentheses (,).

A *term* in L is defined to be an element of the free group freely generated by the variables $x_0, x_1, x_2, \ldots$.

A *formula* in L is defined inductively as follows:

(i) If t is any term in L, then $(t = 1)$ is a formula in L.

(ii) If P and Q are formulae in L, then so are $(P \,\&\, Q)$, $(P \text{ or } Q)$, and $(\neg P)$.

(iii) If P is a formula in L, and if x_i is and variable, then $((\forall x_i)P)$ and $((\exists x_i)P)$ are formulae in L.

Nothing is a formula in L unless (i), (ii), and (iii) imply that it is.

We adopt the usual convention of suppressing parentheses in formulae when they are essentially redundant, e.g. we will write $t = 1$, $P \,\&\, Q$, $\neg P$,

$(\forall x_i)P$, and so on. Also, although $\neq$ is not formally in L, we will always write $t \neq 1$ instead of $(\neg(t=1))$ or $\neg(t=1)$.

We define a *free occurrence* of a variable x_i in a formula of L as follows:

If the formula is of the form $t=1$ or $t \neq 1$, then any occurrence of x_i is a free occurrence.

Any free occurrence of x_i in P is also a free occurrence of x_i in $(P \,\&\, Q)$, $(P \text{ or } Q)$, $\neg P$, and, if $i \neq j$, in $(\forall x_j)P$ and $(\exists x_j)P$.

Conversely, any occurrences of x_i in $(\forall x_i)P$ and $(\exists x_i)P$ are called *bounded occurrences*. And a bounded occurrence of x_i in P is also a bounded occurrence of x_i in $(P \,\&\, Q)$, $(P \text{ or } Q)$, $\neg P$, $(\forall x_j)P$, and $(\exists x_j)P$.

Clearly, every occurrence of every variable in every formula of L is either free or else bounded.

A *sentence* in L is a formula in L containing no free occurrences of any variable.

We will interpret the formulae in L as statements about groups. In this interpretation, the variables of L will range over the elements of the group, the functions $\cdot$ and $^{-1}$ will be the group operation and inversion respectively, and the constant symbol 1 will correspond to the identity of the group.

We can extend L to include one (or more) *constants*. (Or, equivalently, we may nominate one of the variables of L to be a constant.) If a is the constant, we write the extension as L(a). Constants are treated in the same way as variables, except that they cannot be quantified over. So the effect of introducing constants, as distinct from variables, is that sentences in L(a) can contain unbounded occurrences of constants.

Let G be a group, and let $\bar{x}_0, \bar{x}_1, \bar{x}_2, \ldots$ be any sequence of elements of G. If P is a formula of L, then we can attach to P a *truth value*, which is its *truth value at* $(G, \bar{x}_0, \bar{x}_1, \ldots)$, by induction, as follows.

The term $t=1$, where $t = w(x_0, x_1, \ldots)$, is said to be true at $(G, \bar{x}_0, \bar{x}_1, \ldots)$ if $w(\bar{x}_0, \bar{x}_1, \ldots) = 1$ is a relation in G, and is said to be false at $(G, \bar{x}_0, \bar{x}_1, \ldots)$ otherwise.

The term $t \neq 1$ is said to be true at $(G, \bar{x}_0, \bar{x}_1, \ldots)$ if $w(\bar{x}_0, \bar{x}_1, \ldots) \neq 1$ holds in G, and false otherwise.

Then $(P \,\&\, Q)$ is true at $(G, \bar{x}_0, \bar{x}_1, \ldots)$ if and only if both P and Q are true, and is false otherwise.

$(P \text{ or } Q)$ is true at $(G, \bar{x}_0, \bar{x}_1, \ldots)$ if either P or Q is true, and is false otherwise.

$\neg P$ is true at $(G, \bar{x}_0, \bar{x}_1, \ldots)$ if and only if P is false.

$(\exists x_i)P$ is true at $(G, \bar{x}_0, \bar{x}_1, \ldots)$ if, for some $\bar{x} \in G$, the formula P is true at $(G, \bar{x}_0, \ldots, \bar{x}_{i-1}, \bar{x}, \bar{x}_{i+1}, \ldots)$, and is false otherwise; $(\forall x_i)P$ is true at $(G, \bar{x}_0, \bar{x}_1, \ldots)$ if and only if P is true at $(G, \bar{x}_0, \ldots, \bar{x}_{i-1}, y, \bar{x}_{i+1}, \ldots)$, for all $y \in G$.

For a language L(a) with a constant, we first pick an element $\bar{a} \in G$

which is to correspond to a, and then we attach to a formula P of $\mathrm{L}(a)$, a true value at $(G, \bar{a}, \bar{x}_0, \bar{x}_1, \ldots)$ in the same fashion.

If the formula P has free occurrences only of $x_{i_1}, \ldots x_{i_k}$, then we only need to know G and $\bar{x}_{i_1}, \ldots, \bar{x}_{i_k}$ (and the constant $\bar{a}$ if $P \in \mathrm{L}(a)$) to determine the truth value of P at $(G, \bar{x}_0, \bar{x}_1, \ldots)$. If P is a sentence in L (or $\mathrm{L}(a)$), then the truth value of P depends on the choice of G (or G and $\bar{a}$) alone. If $P(x_0, \ldots, x_n)$ is a quantifier-free formula, then the truth value of P at $(G, \bar{x}_0, \bar{x}_1, \ldots)$ depends only on $\mathrm{Rel}\,(\bar{x}_0, \ldots, \bar{x}_n)$. So, if P is true at $(G, \bar{x}_0, \ldots, \bar{x}_n)$, then P is true at any $(H, \bar{z}_0, \ldots, \bar{z}_n)$ such that $\mathrm{typ}\,(\bar{z}_0, \ldots, \bar{z}_n) = \mathrm{typ}\,(\bar{x}_0, \ldots, \bar{x}_n)$. Thus, in this case, if $\bar{x}_0, \ldots, \bar{x}_n$ are group elements, it makes sense to say that P is true at $\bar{x}_0, \ldots, \bar{x}_n$, or that $P(\bar{x}_0, \ldots, \bar{x}_n)$ is true. For, this means that P is true at $(G, \bar{x}_0, \ldots, \bar{x}_n)$, where G is any group containing $\langle \bar{x}_0, \ldots, \bar{x}_n \rangle$.

If Δ is a set of sentences in L, then a *model* for Δ is a group G in which all the sentences in Δ are true. (Of course, Δ may have no model, for example, any set containing both $(\forall x)(x = 1)$ and $(\exists x)(x \neq 1)$ has no model.) If Δ is a set of sentences in $\mathrm{L}(a)$, then a model for Δ is a group G, together with a specified element $\bar{a} \in G$, such that every sentence in Δ is true at $(G, \bar{a})$. In this case, we write $\{G, \bar{a}\}$ for the model.

Definition *The (first-order) theory* of a group G is the set of those sentences in L that are true in G.

We can consider any formal language by looking at its syntax, i.e. its structure, by looking at its semantics, i.e. the truth and falsity of its statements within the intended interpretations of the language, or by looking at its logic, i.e. the deductive systems and notions of proof which can be put on to it. It is always necessary, to some extent, to consider the syntax of a language, but we will be considering L from the point of view of its semantics, rather than its logic. So we are interested in group-theoretic models for formulae in L, and the truth of sentences within these models. We shall make frequent use of the following theorem, which is stated without proof.

Gödel's Compactness Theorem *If L is any first-order language, with or without constants, and if Δ is any set of sentences in L, then Δ has a model if and only if every finite subset of Δ has a model.*

9.2 EXISTENTIALLY CLOSED GROUPS AND FIRST-ORDER THEORY

In this section we prove two results that show a 'negative' relationship between existentially closed groups and first-order logic. The first theorem below shows that the property of being existentially closed is not first-order, i.e. there is no $P \in \mathrm{L}$ such that P is true in G if and only if G is existentially closed. We also show that there is a class $\mathscr{C}$ of existentially closed groups, and a formula $P \in \mathrm{L}$, such that P is true at $(M, \bar{x}_0, \ldots, \bar{x}_n)$ (with $M \in \mathscr{C}$) if and only if $\langle \bar{x}_0, \ldots, \bar{x}_n \rangle \subseteq M$ is finite. No such P exists if $\mathscr{C}$ is taken to be an elementary class of groups, so there is no elementary class of existentially closed groups, and there is a first-order formula which shows this.

Theorem 9.1 *If M is an existentially closed group, then there exists a group, with the same first-order theory as M, that is not existentially closed.*

The following lemma is sufficient to prove Theorem 9.1.

Lemma 9.2 *Let X be a group which contains elements of arbitrarily large finite order. Then there exists a group, with the same first-order theory as X, in which not all the elements of infinite order are conjugate.*

Proof We consider the language $\mathrm{L}(a, b)$, where a and b are constants. Consider the sets

$$\Delta = (\text{theory of } X) \cup \{a \neq 1, \quad a^2 \neq 1, \ldots, b \neq 1, \quad b^2 \neq 1, \ldots, \quad (\forall x)(x^{-1}ax \neq b)\}$$

and

$$\Delta_n = (\text{theory of } X) \cup \{a \neq 1, \ldots, a^n \neq 1, \quad b \neq 1, \ldots, b^n \neq 1, \quad (\forall x)(x^{-1}ax \neq b)\}$$

of sentences in $\mathrm{L}(a, b)$. We can choose $\bar{a}_n \in X$ to be an element which has finite order greater than n and we can choose $\bar{b}_n \in X$ to be an element has finite order greater than both n and the order of $\bar{a}_n$. Then $\{X, (\bar{a}_n, \bar{b}_n)\}$ is a model for Δ_n. Every finite subset of Δ belongs to some Δ_n, so every finite subset has a model $\{X, (\bar{a}_n, \bar{b}_n)\}$. Thus, by Gödel's Compactness Theorem, Δ has a model, say $\{Y, (\bar{a}, \bar{b})\}$. Then $\bar{a}$ and $\bar{b}$ have infinite order, but are not conjugate in Y, since $(\forall x)(x^{-1}ax \neq b)$ is true at $\{Y, (\bar{a}, \bar{b})\}$.

Since X is a group, we have, if P is not in the theory of X, that $\neg P$ is in the theory of X. So $\neg P \in \Delta$ and P cannot lie in the theory of Y. Thus, the theory of Y is precisely the theory of X, as required. □

Observe that, for any existentially closed group M, if G is a finitely generated subgroup of M, then $G = \mathrm{C}_M(\mathrm{C}_M(G))$. So the solution of the following exercise, which is left to the reader, is also sufficient to prove Theorem 9.1.

Exercise Let X be a group with a finitely generated infinite subgroup. Show that there exists a group Y, with the same theory as X, having a finitely generated subgroup G such that $G \neq C_Y(C_Y(G))$.

Definition An *elementary class* $\mathscr{C}$ of groups is the collection of all groups that are models for some set Δ of sentences in L. The class $\mathscr{C}$ is uniquely determined by Δ. (Clearly, an elementary class is isomorphism-closed, so this definition of a type of class is consistent with the definition of a class used in Chapter 3.)

We shall write $s\mathscr{C}$ for the class of all groups which are subgroups of members of $\mathscr{C}$.

Theorem 9.1 shows that no non-empty elementary class can consist only of existentially closed groups. The next theorem gives a first-order formula which distinguishes between elementary classes and classes of existentially closed groups.

Theorem 9.3 (i) *If $\mathscr{C}$ is an elementary class of groups such that $s\mathscr{C}$ contains infinitely many types of k-generator finite groups, then there is no formula $P(x_1, \dots, x_k) \in$ L such that for all $G \in \mathscr{C}$:*

$P(x_1, \dots, x_k)$ *is true at $(G, \bar{x}_1, \dots, \bar{x}_k)$ if and only if $\langle \bar{x}_1, \dots, \bar{x}_k \rangle$ is finite.*

(ii) *If we replace $\mathscr{C}$, in* (i), *by a class of existentially closed groups, then there is such a formula.*

Proof (i) We show that, if $P(x_1, \dots, x_k)$ is true at $(G, \bar{x}_1, \dots, \bar{x}_k)$, for all $G \in \mathscr{C}$ and for all $\bar{x}_i$ $(1 \leq i \leq k)$ such that $\langle \bar{x}_1, \dots, \bar{x}_k \rangle$ is finite, then there is a group $H \in \mathscr{C}$ such tht P is true at $(H, h_1, \dots, h_k)$, but $\langle h_1, \dots, h_k \rangle$ is infinite. This is clearly sufficient to prove (i).

If $\langle \bar{x}_1, \dots, \bar{x}_k \rangle$ is finite, then it has a finite set of defining relations and, up to isomorphism, only finitely many homomorphic images. Thus there is a formula $Q(x_1, \dots, x_k)$ such that, if $g_1, \dots, g_k$ satisfy Q, then

$$\text{typ}(\bar{x}_1, \dots, \bar{x}_k) = \text{typ}(g_1, \dots, g_k).$$

We take one such formula Q_i for each isomorphism type in $s\mathscr{C}$.

Let Δ be the set of sentences which define $\mathscr{C}$, and suppose that P is a formula which is true at $(G, \bar{x}_1, \dots, \bar{x}_k)$ whenever $G \in \mathscr{C}$ and $\langle \bar{x}_1, \dots, \bar{x}_k \rangle$ is finite. We consider the set

$$\Gamma = \Delta \cup \{\neg Q_i(a_1, \dots, a_k) \mid i \in \mathbb{N}\} \cup \{P(a_1, \dots, a_k)\}$$

of sentences in $L(a_1, \dots, a_k)$. Since $\mathscr{C}$ contains infinitely many types of k-generator finite groups, we can find $G \in \mathscr{C}$ and $\{\bar{a}_1, \dots, \bar{a}_k\} \subseteq G$ such that $\langle \bar{a}_1, \dots, \bar{a}_k \rangle$ is finite, but is not of the type defined by Q_i, for any $i \leq n$. Thus, Q_i is not true at $(G, \bar{a}_1, \dots, \bar{a}_k)$ $(i \leq n)$, and so

$\{G, (\bar{a}_1, \dots, \bar{a}_k)\}$ is a model for the set

$$\Gamma_n = \Delta \cup \{\neg Q_i(a_1, \dots, a_k) \mid i \leq n\} \cup \{P(a_1, \dots, a_k)\}.$$

Since every finite subset of Γ lies in some Γ_n, and therefore has a model, Γ has a model, $\{H, (\bar{a}_1^*, \dots, \bar{a}_k^*)\}$ (say). Let $h_i = \bar{a}_i^*$ $(1 \leq i \leq k)$. Since Δ is true in H, we have that $H \in \mathscr{C}$, and P is true at $(H, h_1, \dots, h_k)$. But $(h_1, \dots, h_k)$ satisfies $\neg Q_i$, for all $i \in \mathbb{N}$, so $\langle h_1, \dots, h_k \rangle$ must be infinite. Thus there is no formula $P(x_1, \dots, x_k)$ of the type in the statement of the theorem.

(ii) In order to construct P, we need some discussion of the group Alt $\mathbf{\Omega}$, the group of all even permutations of the set $\mathbb{Z} \cup \{\infty\} = \mathbf{\Omega}$ which have finite support. (Recall that a function f has finite support if the set $\{x \mid f(x) \neq x\}$ is finite.) We work within the group $S_{\mathbf{\Omega}}$ of all permutations on $\mathbf{\Omega}$. Recall that Alt $\mathbf{\Omega}$ is generated by the 3-cycles in $S_{\mathbf{\Omega}}$.

Let $g \in S_{\mathbf{\Omega}}$ be the permutation which maps z to $z+1$, for all $z \in \mathbb{Z}$, and which fixes ∞. Let $h = (0\ \infty)$. Then $g^{-1}h^{-1}gh = (1\ 0\ \infty)$, and we see, by conjugating and taking products, that the derived subgroup $\langle g, h \rangle' \subseteq S_{\mathbf{\Omega}}$ contains all 3-cycles. Thus Alt $\mathbf{\Omega} \subseteq \langle g, h \rangle'$. For each $q \in \langle g, h \rangle$, we have

$$q = g^{i_1}hg^{i_2}h \cdots g^{i_n}.$$

Then, for $m \in \mathbf{\Omega}$, if

$$m \notin \{-i_1, -(i_1+i_2), \dots, -(i_1+\cdots+i_{n-1}), \infty\},$$

we have that $mq = m + (i_1 + \cdots + i_n)$. So, given $q \in \langle g, h \rangle$, there exists an integer n_q such that, for all but finitely many $m \in \mathbf{\Omega}$, the equation $mq = m + n_q$ holds. Thus, if $p \in \langle g, h \rangle$, then, for all but finitely many $m \in \mathbf{\Omega}$, we have

$$m[p, q] = (m - n_p)q^{-1}pq = (m - n_p - n_q)pq = m.$$

So $[p, q]$ has finite support. All commutators are even permutations, so $[p, q] \in$ Alt $\mathbf{\Omega}$. Thus Alt $\mathbf{\Omega} = \langle g, h \rangle'$. Since Alt $\mathbf{\Omega}$ is simple and non-Abelian, it is perfect. Any finite group is embeddable in Alt $\mathbf{\Omega}$, and, since every element of Alt $\mathbf{\Omega}$ has finite support, it follows that every finitely generated subgroup of Alt $\mathbf{\Omega}$ is finite. Thus, a finitely generated group $\langle a_1, \dots, a_k \rangle$ is finite if and only if there exist elements $y_1, \dots, y_k, z_1, \dots, z_k$ of $\langle g, h \rangle$ such that the map $a_i \mapsto y_i^{-1}z_i^{-1}y_iz_i$ $(1 \leq i \leq k)$ extends to a monomorphism $\langle a_1, \dots, a_k \rangle \to \langle g, h \rangle$.

Now, it is easy to see that $\langle g, h \rangle$ has solvable word problem, as follows. If

$$q = g^{i_1}hg^{i_2}h \cdots g^{i_n},$$

then $q = 1$ implies that $mq = m$, for all $m \in \mathbf{\Omega}$, and hence that $i_1 + i_2 + \cdots + i_n = 0$. If $i_1 + i_2 + \cdots + i_n = 0$, then $q = 1$ if and only if q fixes the set $\{-i_1, \dots, -(i_1 + \cdots + i_{n-1}), \ \infty\}$ pointwise, and this is a finite condition that can be checked mechanically. So, there exists a finitely presented group

$$H = \langle u_1, \ldots, u_s \mid r_j(\boldsymbol{u}) = 1, \quad 1 \leqslant j \leqslant m \rangle,$$

and elements γ and η of H, such that the map

$$g \mapsto \gamma, \qquad h \mapsto \eta$$

extends to an embedding $\langle g, h \rangle \to H$. We take $P(x_1, \ldots, x_k)$ to be

$$\begin{aligned}(\exists u_1, \ldots, u_k, y_1, \ldots, y_k, z_1, \ldots, z_k)(\forall t)\big(r_1(\boldsymbol{u}) = 1 \,\&\, \ldots \,\&\, r_m(\boldsymbol{u}) = 1 \\ \&\, x_1 = [y_1, z_1] \,\&\, \ldots \,\&\, x_k = [y_k, z_k] \,\&\, ((t^{-1}gt = g \,\&\, t^{-1}ht = h) \Rightarrow \\ (t^{-1}y_1t = y_1 \,\&\, t^{-1}z_1t = z_1 \,\&\, \ldots \,\&\, t^{-1}y_kt = y_k \,\&\, t^{-1}z_kt = z_k))\big).\end{aligned}$$

Note that the symbol $\Rightarrow$ is not formally in L, but we write $A \Rightarrow B$ for $\neg(A \,\&\, \neg B)$.

We want to show that, if M is an existentially closed group, then P is true at $(M, \bar{x}_1, \ldots, \bar{x}_k)$ if and only if $\langle \bar{x}_1, \ldots, \bar{x}_k \rangle$ is finite.

If P is true at $(M, \bar{x}_1, \ldots, \bar{x}_k)$, let $\bar{u}_1, \ldots, \bar{u}_s, \bar{y}_1, \ldots, \bar{y}_k, \bar{z}_1, \ldots, \bar{z}_k$ be the elements of M whose existence is required by the truth of $P(\bar{x}_1, \ldots, \bar{x}_k)$, and let

$$\bar{\gamma} = \gamma(\bar{u}_1, \ldots, \bar{u}_s), \qquad \bar{\eta} = \eta(\bar{u}_1, \ldots, \bar{u}_s).$$

So $\langle \bar{u}_1, \ldots, \bar{u}_s \rangle$ is a homomorphic image of H, since $\bar{r}_i = 1$ holds in M, for $1 \leqslant i \leqslant m$.

Now, if $\bar{y}_i \notin \langle \bar{\gamma}, \bar{\eta} \rangle$, then, in some HNN-extension of M, we can solve the set

$$\{t^{-1}\bar{\gamma}t = \bar{\gamma}, \quad t^{-1}\bar{\eta}t = \bar{\eta}, \quad t^{-1}\bar{y}_it \neq \bar{y}_i\}.$$

So we could choose such a t in M; this contradicts P, which contains the statement

$$(\forall t)\big(((t^{-1}\bar{\eta}t = \bar{\eta}) \,\&\, (t^{-1}\bar{\gamma}y = \bar{\gamma})) \Rightarrow t^{-1}\bar{y}_it = \bar{y}_i\big).$$

Thus $\bar{y}_i$ and $\bar{z}_i$ belong to $\langle \bar{\gamma}, \bar{\eta} \rangle$, and so $\bar{x}_i \in \langle \bar{\gamma}, \bar{\eta} \rangle'$. So $\langle \bar{x}_1, \ldots, \bar{x}_k \rangle$ is a homomorphic image of some group $\langle \alpha_1, \ldots, \alpha_k \rangle \subseteq \langle g, h \rangle'$. Since $\langle \alpha_1, \ldots, \alpha_k \rangle$ is finitely generated, and therefore finite, $\langle \bar{x}_1, \ldots, \bar{x}_k \rangle$ is finite.

Conversely, if $\langle \bar{x}_1, \ldots, \bar{x}_k \rangle \subseteq M$ is finite, then $\langle \bar{x}_1, \ldots, \bar{x}_k \rangle$ can be embedded in $\langle g, h \rangle' \subseteq H$. So, the equations

$$\{r_j(\boldsymbol{u}) = 1, \quad \bar{x}_i = y_i^{-1}z_i^{-1}y_iz_i, \quad y_i = w_i(\gamma, \eta), z_i = v_i(\gamma, \eta) \mid \\ 1 \leqslant j \leqslant m, \quad 1 \leqslant i \leqslant k\} = \mathcal{S}$$

can be solved in the free product of M with H, amalgamating $\langle \bar{x}_1, \ldots, \bar{x}_k \rangle$. Thus, we can find solutions $\bar{u}_1, \ldots, \bar{u}_s, \bar{y}_1, \ldots, \bar{y}_k, \bar{z}_1, \ldots, \bar{z}_k$ for $\mathcal{S}$ in M. So P is true at $(M, \bar{x}_1, \ldots, \bar{x}_k)$, and we have the result. □

In the above proof, the formula P has not been given explicitly, since The Subgroup Theorem does not give η and γ explicitly. So we now show how to embed $\langle g, h\rangle$ in a finitely presented group directly, without using The Subgroup Theorem.

Continuing with the notation as in the proof of Theorem 9.3(ii), let $h_i = g^{-i}hg^i = (i\ \ \infty)$, $i \in \mathbb{Z}$. Then

$$K = \langle h_i \mid i \in \mathbb{Z}\rangle$$

is exactly the set of permutations of $\boldsymbol{\Omega}$ with finite support, and K has the set

$$\{h_i^2 = 1,\quad (h_ih_j)^3 = 1,\quad (h_ih_jh_kh_j)^2 = 1 \mid i<j<k;\ \ i, j, k \in \mathbb{Z}\}$$

as a set of defining relations. Also, $G = \langle g, h\rangle$ has, as a set of defining relations, the above set together with the relations $g^{-1}h_ig = h_{i+1}$ $(i \in \mathbb{Z})$. There is an isomorphism θ from $G_1 = \langle g^3, h_i \mid i \in \mathbb{Z}\rangle$ to $G_2 = \langle g^4, h_i \mid i \in \mathbb{Z},\ \ i \not\equiv 2 \pmod 4\rangle$, given by

$$g^3\theta = g^4, \qquad (h_{3i+\alpha})\theta = h_{4i+\alpha}, \qquad\qquad (\alpha \in \{-1, 0, 1\},\quad i \in \mathbb{Z}).$$

So we can form the HNN-extension

$$H = \langle G, t \mid t^{-1}g^3t = g^4,\quad t^{-1}h_it = h_i\theta\rangle.$$

In fact, as defining relations for $H = \langle g, h, t\rangle$, we can take the set

$$\{h^2 = 1,\quad t^{-1}g^3t = g^4,\quad t^{-1}g^{-1}hgt = g^{-1}hg,\quad (ghg^{-1}hghg^{-1}h)^2 = 1,$$
$$t^{-1}ht = h,\quad t^{-1}ghg^{-1}t = ghg^{-1},\quad (hg^{-1}hg)^3 = 1\}.$$

So we have an explicit embedding of $\langle g, h\rangle$ in a finitely presented group.

9.3 STABLE EVALUATION

In the first part of this section, we define the notion of stable truth of a formula, and use it to describe a whole class of sentences that are true in every existentially closed group. In the second part of the section, we shall give contrasting examples of sentences that are true in some existentially closed groups but false in others.

A $\forall_n$-formula and a $\exists_n$-formula are defined inductively as follows.

If P is a quantifier-free formula, then $(\forall x_1, \ldots, x_r)P$ is a $\forall_1$-formula and $(\exists x_1, \ldots, x_r)P$ is a $\exists_1$-formula.

If P_1 is a $\forall_n$-formula and P_2 is a $\exists_n$-formula, and if the variables $y_1, \ldots, y_s$ are not already governed by quantifiers in P_i, then $(\forall y_1, \ldots, y_s)P_2$ is a $\forall_{n+1}$-formula and $(\exists y_1, \ldots, y_s)P_1$ is a $\exists_{n+1}$-formula.

We allow the possibility that $s = 0$, so that a $\forall_n$-formula is also a $\exists_{n+1}$-formula, and a $\exists_n$-formula is also a $\forall_{n+1}$-formula

A group $\mathcal{G}$ will be said to be *fg-complete* if it satisfies the following two conditions.

(i) Every finitely generated group belongs to Sk $\mathcal{G}$.

(ii) Any two embeddings of a given finitely generated group into $\mathcal{G}$ are conjugate in $\mathcal{G}$.

Let P be a formula in L, with free variables $x_1, \ldots, x_r$. Consider sets $\{g_1, \ldots, g_r\} \subseteq \mathcal{G}$ and $\{h_1, \ldots, h_r\} \subseteq \mathcal{G}_1$, where $\mathcal{G}$ and $\mathcal{G}_1$ are *fg*-complete and typ $(g_1, \ldots, g_r) =$ typ $(h_1, \ldots, h_r)$. We show, by induction on P, that if P is true at $(\mathcal{G}, g_1, \ldots, g_r)$ then P is true at $(\mathcal{G}_1, h_1, \ldots, h_r)$.

If P is quantifier-free, then the truth value of P at $(\mathcal{G}, g_1, \ldots, g_r)$ depends only on Rel $(g_1, \ldots, g_r)$, so the result is clear.

Suppose that $P = (\forall z)Q$, where Q has free variables $x_1, \ldots, x_r$, and z. If P is true at $(\mathcal{G}, g_1, \ldots, g_r)$, then Q is true at $(\mathcal{G}, g_1, \ldots, g_r, g)$, for all $g \in \mathcal{G}$. Now, for all $h \in \mathcal{G}_1$, $\langle h_1, \ldots, h_r, h\rangle \in$ Sk $\mathcal{G}$, by (i). Also, by (ii), we can find some $g \in \mathcal{G}$ such that

$$\text{typ}\,(h_1, \ldots, h_r, h) = \text{typ}\,(g_1, \ldots, g_r, g).$$

By induction, Q is true at $(\mathcal{G}_1, h_1, \ldots, h_r, h)$, so P is true at $(\mathcal{G}_1, h_1, \ldots, h_r)$.

If $P = (\exists z)Q$, then, for some $g \in \mathcal{G}$, Q is true at $(\mathcal{G}, g_1, \ldots, g_r, g)$. So, since there exists $h \in \mathcal{G}$ such that typ $(g_1, \ldots, g_r, g) =$ typ $(h_1, \ldots, h_r, h)$, the formula Q is true at some $(\mathcal{G}_1, h_1, \ldots, h_r, h)$. Thus P is true at $(\mathcal{G}_1, h_1, \ldots, h_r)$.

So we see that, if $\mathcal{G}$ is an *fg*-complete group and if P is any formula with free variables $x_1, \ldots, x_r$, then the truth value of P at $(\mathcal{G}, g_1, \ldots, g_r)$ depends only on typ $(g_1, \ldots, g_r)$. This truth value is called the *stable evaluation* of P at $(g_1, \ldots, g_r)$.

We emphasize that to say 'P is stably true at $(g_1, \ldots, g_n)$' is just to say 'there is some *fg*-complete group $\mathcal{G}$ containing $\langle g_1, \ldots, g_n\rangle$ such that P is true at $(\mathcal{G}, g_1, \ldots, g_n)$'.

We note that, given any group Y, there exists an *fg*-complete group containing Y. To see this, first choose one group from each isomorphism class of finitely generated groups and define $\mathcal{G}_0$ to be the free product of all of these groups and the group Y. Then, for $i \in \mathbb{N}$, if $\mathcal{G}_i$ has been defined, let $\mathcal{G}_{i+1}$ be a repeated HNN-extension of $\mathcal{G}_i$ in which any two isomorphic finitely generated groups in $\mathcal{G}_i$ are conjugate in $\mathcal{G}_{i+1}$. Then $\mathcal{G} = \bigcup_{i \in \mathbb{N}} \mathcal{G}_i$ is the required group.

So, for any set $\{g_1, \ldots, g_r\}$ of elements of some group, the group $\langle g_1, \ldots, g_r\rangle$ lies in some *fg*-complete group. Thus any formula P with free variables $x_1, \ldots, x_r$ has a stable evaluation at every r-tuple $(g_1, \ldots, g_r)$ of group elements. Further, if $P = (\forall z)Q$, where Q has free variables $x_1, \ldots, x_r, z_1, \ldots, z_s$, then P is stably true at $(g_1, \ldots, g_r)$ if and only if Q is stably true at $(g_1, \ldots, g_r, h_1, \ldots, h_s)$, for all sets $\{h_1, \ldots, h_s\}$ of group elements.

We want to show that the class of $\forall_3$-sentences that are stably true is exactly the class of $\forall_3$-sentences which are true in every existentially closed group. To do this, we need the following lemma.

Lemma 9.4 *Let $f = (\forall y_1, \dots, y_s)P$ be a $\forall_1$-formula, with free variables $x_1, \dots, x_r$, that is stably true at $(g_1, \dots, g_r)$. Then there exist finite subsets $A \subseteq \mathrm{Rel}(g_1, \dots, g_r)$ and $B \subseteq \mathrm{Nonrel}(g_1, \dots, g_r)$ such that f is stably true at $(h_1, \dots, h_r)$ whenever $A \subseteq \mathrm{Rel}(h_1, \dots, h_r)$ and $B \subseteq \mathrm{Nonrel}(h_1, \dots, h_r)$.*

Proof Let $\langle k_1, \dots, k_r\rangle$ be any r-generator group. We shall think of $\mathrm{Rel}(k_1, \dots, k_r)$ as a set of words on the variables $x_1, \dots, x_r$ and their inverses. Now, f is stably true at $(k_1, \dots, k_r)$ is and only if, for all fg-complete groups $\mathscr{G}$, and for all $\bar{x}_1, \dots, \bar{x}_r \in \mathscr{G}$ such that

$$\text{type}(k_1, \dots, k_r) = \text{typ}(\bar{x}_1, \dots, \bar{x}_r),$$

f is true at $(\mathscr{G}, \bar{x}_1, \dots, \bar{x}_r)$.

Consider the extended language $L(a_1, \dots, a_r, b_1, \dots, b_s)$, and let $\Delta(\boldsymbol{k})$ be the set of sentences

$$\{w(\boldsymbol{a}) = 1 \mid w \in \mathrm{Rel}(k_1, \dots, k_r)\} \cup \{u(\boldsymbol{a}) \neq 1 \mid u \in \mathrm{Nonrel}(k_1, \dots, k_r)\} \cup \{\neg P(\boldsymbol{a}, \boldsymbol{b})\}.$$

To prove the result, we will show that f is stably true at $(k_1, \dots, k_r)$ if and only if $\Delta(\boldsymbol{k})$ has no models. Once this is established, we will have, since f is stably true at $(g_1, \dots, g_r)$, that $\Delta(\boldsymbol{g})$ has no models, and so, by Gödel's Theorem, some finite subset $\Gamma(\boldsymbol{g})$ of $\Delta(\boldsymbol{g})$ has no models. We can then take

$$A = \{w \in \mathrm{Rel}(g_1, \dots, g_s) \mid w(\boldsymbol{a}) = 1 \in \Gamma(\boldsymbol{g})\},$$
$$B = \{u \in \mathrm{Nonrel}(g_1, \dots, g_s) \mid u(\boldsymbol{a}) \neq 1 \in \Gamma(\boldsymbol{g})\},$$

because, if

$$A \subseteq \mathrm{Rel}(h_1, \dots, h_r) \quad \text{and} \quad B \subseteq \mathrm{Nonrel}(h_1, \dots, h_r),$$

then

$$\Gamma(\boldsymbol{g}) \subseteq \Delta(\boldsymbol{h}),$$

so $\Delta(\boldsymbol{h})$ has no models, and hence f is stably true at $(h_1, \dots, h_r)$, as required.

Suppose that

$$\{G, (\bar{a}_1, \dots, \bar{a}_r, \bar{b}_1, \dots, \bar{b}_s)\}$$

is a model for $\Delta(\boldsymbol{k})$, so that

$$\text{typ}(\bar{a}_1, \dots, \bar{a}_r) = \text{typ}(k_1, \dots, k_r).$$

Embed G in an fg-complete group $\mathscr{G}$, and let $\bar{a}_i^*$ and $\bar{b}_i^*$ be the images of

$\bar{a}_i$ and $\bar{b}_j$ $(1 \leq i \leq r,\ 1 \leq j \leq s)$ under this embedding. Then, since $\neg P$ is quantifier free, $\neg P$ must be true at $(\mathscr{G}, \bar{a}_1^*, \ldots, \bar{a}_r^*, \bar{b}_1^*, \ldots, \bar{b}_s^*)$. Thus f is false at $(\mathscr{G}, \bar{a}_1^*, \ldots, \bar{a}_r^*)$, with

$$\text{typ}\,(\bar{a}_1^*, \ldots, \bar{a}_r^*) = \text{typ}\,(k_1, \ldots, k_r),$$

and so f is not stably true at $(k_1, \ldots, k_r)$.

Now suppose that $\Delta(\boldsymbol{k})$ has no models. Let $\mathscr{G}$ be any fg-complete group, and let $\bar{a}_1, \ldots, \bar{a}_r$ be elements of $\mathscr{G}$ such that

$$\text{typ}\,(\bar{a}_1, \ldots, \bar{a}_r) = \text{typ}\,(k_1, \ldots, k_r).$$

Then $\{\mathscr{G}, (\bar{a}_1, \ldots, \bar{a}_r)\}$ is a model for the set

$$\Delta(\boldsymbol{k})\backslash\{\neg P(\boldsymbol{a}, \boldsymbol{b})\}.$$

which lies in $\mathrm{L}(a_1, \ldots, a_r)$. Thus, for each sequence $(\bar{b}_1, \ldots, \bar{b}_s)$ of elements of $\mathscr{G}$, since $\{\mathscr{G}, (\bar{a}_1, \ldots, \bar{a}_r, \bar{b}_1, \ldots, \bar{b}_s)\}$ cannot be a model for $\Delta(\boldsymbol{k})$, we must have that $\neg P(\boldsymbol{a}, \boldsymbol{b})$ is false at $(\mathscr{G}, \bar{a}_1, \ldots, \bar{a}_r, \bar{b}_1, \ldots, \bar{b}_s)$. Thus, P is true at $(\mathscr{G}, \bar{a}_1, \ldots, \bar{a}_r, \bar{b}_1, \ldots, \bar{b}_s)$, for all $\bar{b}_1, \ldots, \bar{b}_s$, and so f is true at $(\mathscr{G}, \bar{a}_1, \ldots, \bar{a}_r)$. Thus, f is stably true at $(k_1, \ldots, k_r)$ if and only if $\Delta(\boldsymbol{k})$ has no models. □

Theorem 9.5 *A $\forall_3$-sentence of* L *is true in every existentially closed group if and only if it is stably true.*

Proof Any sentence is either true or false in any given group. Thus a sentence is stably true if and only if it is true in some, and therefore (as in the discussion before the definition) every, fg-complete group. To prove that every sentence that is true in every existentially closed group is actually stably true, it is sufficient to show that every fg-complete group is existentially closed. (In fact it would be sufficient to show that some fg-complete group is existentially closed.)

Let $\mathscr{G}$ be an fg-complete group, and let $\mathscr{S}$ be a finite set of equations and inequalities defined, and soluble, over $\mathscr{G}$. Then $\mathscr{S}$ is defined over a finitely generated subgroup G of $\mathscr{G}$, and is soluble in a finitely generated group $H \supseteq G$. Let $\alpha: H \to \mathscr{G}$ be an embedding. Then G and $G\alpha$ are conjugate in $\mathscr{G}$, and so we may take $\alpha|_G$ to be the identity. Thus $\mathscr{S}$ is soluble in $H\alpha \subseteq \mathscr{G}$, and hence $\mathscr{G}$ is existentially closed, as required.

Now, suppose that $(\forall\boldsymbol{x})(\exists\boldsymbol{y})(\forall\boldsymbol{z})P$ is a stably true sentence, and let M be an existentially closed group. We want to show that $(\forall\boldsymbol{x})(\exists\boldsymbol{y})(\forall\boldsymbol{z})P$ is true in M.

Since $(\forall\boldsymbol{x})(\exists\boldsymbol{y})(\forall\boldsymbol{z})P$ is stably true, it is true in any fg-complete group. So, for any $\{\bar{x}_1, \ldots, \bar{x}_n\} \subseteq M$, and for any fg-complete group $\mathscr{G}$ containing M, the sentence $(\exists\boldsymbol{y})(\forall\boldsymbol{z})P$ is true at $(\mathscr{G}, \bar{x}_1, \ldots, \bar{x}_n)$, i.e. since such a group $\mathscr{G}$ exists, $(\exists\boldsymbol{y})(\forall\boldsymbol{z})P$ is stably true at $(\bar{x}_1, \ldots, \bar{x}_n)$. Thus, there exist $\bar{y}_1, \ldots, \bar{y}_m$, in $\mathscr{G}$, such that $(\forall\boldsymbol{z})P$ is true at

$$(\mathscr{G}, \bar{x}_1, \ldots, \bar{x}_n, \bar{y}_1, \ldots, \bar{y}_m).$$

So $(\forall \boldsymbol{z})P$ is stably true at $(\bar{x}_1, \ldots, \bar{x}_n, \bar{y}_1, \ldots, \bar{y}_m)$. We want to show that we can take $\bar{y}_1, \ldots, \bar{y}_m$ to lie in M. For then $(\forall \boldsymbol{z})P$ is true at $(M, \bar{x}_1, \ldots, \bar{y}_m)$, since $M \subseteq \mathscr{G}$, and so $(\forall \boldsymbol{x})(\exists \boldsymbol{y})(\forall \boldsymbol{z})P$ is true in M, as required.

Suppose, then, that we have $\bar{y}_1^*, \ldots, \bar{y}_m^*$ in $\mathscr{G}$ such that $(\forall \boldsymbol{z})P$ is stably true at $(\bar{x}_1, \ldots, \bar{x}_n, \bar{y}_1^*, \ldots, \bar{y}_m^*)$. By Lemma 9.4, there exist finite sets $A \subseteq \mathrm{Rel}\,(\bar{x}_1, \ldots, \bar{y}_m^*)$ and $B \subseteq \mathrm{Nonrel}\,(\bar{x}_1, \ldots, \bar{y}_m^*)$, such that for any $\bar{y}_1, \ldots, \bar{y}_m$ in any group, if $A \subseteq \mathrm{Rel}\,(\bar{x}_1, \ldots, \bar{y}_m)$ and if $B \subseteq \mathrm{Nonrel}\,(\bar{x}_1, \ldots, \bar{y}_m)$, then $(\forall \boldsymbol{z})P$ is stably true at $(\bar{x}_1, \ldots, \bar{y}_m)$. Let

$$A' = \{w(\bar{x}_1, \ldots, \bar{x}_n, y_1, \ldots, y_m) = 1 \mid w(\bar{x}_1, \ldots, \bar{y}_m^*) \in A\},$$
$$B' = \{w(\bar{x}_1, \ldots, \bar{x}_n, y_1, \ldots, y_m) \neq 1 \mid w(\bar{x}_1, \ldots, \bar{y}_m^*) \in B\}.$$

be finite sets of equations and inequalities, in the variables $y_1, \ldots, y_m$, defined over M. The set $A' \cup B'$ is soluble over M, since $\{\bar{y}_1^*, \ldots, \bar{y}_m^*\}$ is a set of solutions in $\mathscr{G} \supseteq M$. So $A' \cup B'$ is soluble in M, and we let $\bar{y}_1, \ldots, \bar{y}_m \in M$ be solutions. By construction, $A \subseteq \mathrm{Rel}\,(\bar{x}_1, \ldots, \bar{y}_m)$ and $B \subseteq \mathrm{Nonrel}\,(\bar{x}_1, \ldots, \bar{y}_m)$, and so $(\forall \boldsymbol{z})P$ is stably true at $(\bar{x}_1, \ldots, \bar{y}_m)$. Thus $(\forall \boldsymbol{z})P$ is true at

$$(\mathscr{G}, \bar{x}_1, \ldots, \bar{x}_n, \bar{y}_1, \ldots, \bar{y}_m),$$

where $\bar{y}_1, \ldots, \bar{y}_m$ lie in M, as required. □

Corollary 9.6 *A $\forall_2$-sentence (and a $\exists_2$-sentence) has the same truth value in every existentially closed group.*

Proof The negation of a $\forall_2$-sentence $f = (\forall \boldsymbol{x})(\exists \boldsymbol{y})P$ is a $\exists_2$-sentence $\neg f = (\exists \boldsymbol{x})(\forall \boldsymbol{y})(\neg P)$. Given an *fg*-complete group $\mathscr{G}$, either f or $\neg f$ is true in $\mathscr{G}$. Thus, either f or $\neg f$ is stably true. Both f and $\neg f$ are $\forall_3$-sentences and so, by Theorem 9.5, whichever is true in $\mathscr{G}$ is true in every existentially closed group. □

We consider now some examples of sentences that are true in some existentially closed groups, but false in others.

Example 1

$$(\exists x, y)(\forall u, v)(\exists s)(\forall t)\big((t^{-1}xt = x \quad \& \quad t^{-1}yt = y) \Rightarrow$$
$$(t^{-1}s^{-1}ust = s^{-1}us \quad \& \quad t^{-1}s^{-1}vst = s^{-1}vs)\big). \quad (1)$$

Clearly, for any group G, if there exist x and y in G such that, for all $u,v \in G$, we can choose $s \in G$ such that $s^{-1}\langle u, v\rangle s \subseteq \langle x, y\rangle$, then

$$\mathrm{C}_G(\langle x, y\rangle) \subseteq \mathrm{C}_G(s^{-1}\langle u, v\rangle s),$$

and (1) holds in G. If (1) holds in G, then there exist $x,y \in G$ such that

$$\mathrm{C}_G(\langle x,y\rangle) \subseteq \mathrm{C}_G(s^{-1}\langle u, v\rangle s).$$

If $s^{-1}\langle u, v\rangle s \nsubseteq \langle x, y\rangle$, then we can find $w \in \langle u, v\rangle$ such that $s^{-1}ws \notin \langle x, y\rangle$. The equations and inequalities

$$\{t^{-1}xt = x, \quad t^{-1}yt = y, \quad t^{-1}s^{-1}wst \neq s^{-1}ws\}$$

are then soluble in the HNN-extension $\langle G, t \mid t^{-1}xt = x, \quad t^{-1}yt = y\rangle$. So, if G is existentially closed, then we can find $t \in G$ such that $t \in \mathrm{C}_G(\langle x, y\rangle)$ but $t \notin \mathrm{C}_G(s^{-1}\langle u, v\rangle s)$, which is a contradiction if (1) holds. Thus we have that (1) is true in an existentially closed group M if and only if there is a group $\langle x, y\rangle$ in M such that every group $\langle u, v\rangle \subseteq M$ is conjugate to a subgroup of $\langle x, y\rangle$.

By Theorem 1.12, we see that every finitely generated subgroup of M can be embedded in a two-generator subgroup of M. Embeddings within existentially closed groups can always be realized as conjugacy within the group; so, if (1) is true in M, then every finitely generated subgroup of M can be embedded in $\langle x, y\rangle$. Thus (1) is true in M if and only if M contains a finitely generated subgroup G (which can, by Theorem 1.12, be taken to be 2-generator) such that $\mathrm{Sk}\, M = \mathrm{Sk}\, G$.

We have already seen, in Corollary 6.4, that this holds for any group M_X ($X \subseteq \mathbb{N}$). But there exist existentially closed groups for which (1) is false. For instance, let $M = \bigcup_{i\in\mathbb{N}} M_{X_i}$, where the set $\{X_i \mid i \in \mathbb{N}\}$ is chosen so that

$$X_i \leqslant_{\mathrm{e}} X_{i+1} \quad \text{but} \quad X_{i+1} \nleqslant_{\mathrm{e}} X_i \qquad (i \in \mathbb{N}).$$

Then M is certainly existentially closed, in fact, M is the unique countable existentially closed group with the property that $G \in \mathrm{Sk}\, M$ if and only if $\mathrm{Rel}\, G \leqslant_{\mathrm{e}} X_i$, for some $i \in \mathbb{N}$. But (1) cannot be true in M, for the following reason. If G is a finitely generated subgroup of M, then $\mathrm{Rel}\, G \leqslant_{\mathrm{e}} X_i$, for some i, but we can find a finitely generated group H such that

$$\mathrm{Rel}\, H \equiv_{\mathrm{e}} X_{i+1} \nleqslant_{\mathrm{e}} \mathrm{Rel}\, G;$$

so $H \in \mathrm{Sk}\, M$ but $H \notin \mathrm{Sk}\, G$, and $\mathrm{Sk}\, M \neq \mathrm{Sk}\, G$.

Example 2 Let $\mathrm{Hom}\,(x_1, \ldots, x_r, y_1, \ldots, y_r, s, t)$ denote the set

$$\{s^{-1}y_isx_j = x_js^{-1}y_is, \quad t^{-1}x_it = x_is^{-1}y_is \mid 1 \leqslant i,j \leqslant r\}.$$

(For notational convenience, if

$$X = \{w_i = 1, \quad u_j \neq 1 \mid 1 \leqslant i \leqslant n, \quad 1 \leqslant j \leqslant m\}$$

is any finite set of equations and inequalities, we will also denote by X the formula

$$(w_1 = 1)\,\&\cdots\&\,(w_n = 1)\,\&\,(u_1 \neq 1)\,\&\cdots\&\,(u_m \neq 1),$$

in L. Of course, $\bar{x}_1, \ldots, \bar{x}_k$ are solutions for X if and only if the formula X is true at $(\bar{x}_1, \ldots, \bar{x}_k)$.)

We recall, from Lemma 5.2, that there is a homomorphism from $\langle g_1, \ldots, g_r\rangle$ to $\langle h_1, \ldots, h_r\rangle$, carrying g_i to h_i $(1 \leq i \leq r)$, if and only if

$$\text{Hom}\,(g_1, \ldots, g_r, h_1, \ldots, h_r, s, t)$$

can be solved in some group containing both $\langle g_1, \ldots, g_r\rangle$ and $\langle h_1, \ldots, h_r\rangle$. So, in an existentially closed group, such a homomorphism exists if and only if

$$(\exists s,t)\,\text{Hom}\,(g_1, \ldots, g_r, h_1, \ldots, h_r, s, t)$$

is true. Let

$$G = \langle g_1, g_2 \mid R(g_1, g_2)\rangle$$

be a finitely presented group, so $R(g_1, g_2)$ is a finite set of relations of the form $r(g_1, g_2) = 1$. We consider the sentence

$$(\exists x_1,x_2)(\forall y_1,y_2)(\exists s,t)\big(R(x_1, x_2)\ \&\ \big(R(y_1, y_2) \Rightarrow \text{Hom}\,(x_1, x_2, y_1, y_2, s, t)\big)\big). \quad (2)$$

We show that (2) is true in an existentially closed group M if and only if $G \in \text{Sk}\,M$.

If $G \in \text{Sk}\,M$, then we can choose k_1 and k_2 to be generators of the image of G in M. Then, if $R(h_1, h_2)$ are true in M for some $h_1,h_2 \in M$, it follows that $\langle h_1, h_2\rangle$ is a homomorphic image of G, and hence of $\langle k_1, k_2\rangle$; thus

$$(\exists s, t)\,\text{Hom}\,(k_1, k_2, h_1, h_2, s, t)$$

is true in M.

If (2) is true in M, and if $k_1,k_2 \in M$ are solutions for (2), suppose that there is a relation $w(k_1, k_2) = 1$ which holds in M, but which is not a consequence of $R(k_1, k_2)$. Then

$$\{R(y_1, y_2), \quad w(y_1, y_2) \neq 1\}$$

is a finite set, which is soluble in $M * G$ and hence in M. If the solutions are $h_1,h_2 \in M$, then there cannot exist a homomorphism from $\langle k_1, k_2\rangle$ to $\langle h_1, h_2\rangle$ that carries k_i to h_i $(i = 1, 2)$. So, if (2) holds in M, then $\langle k_1, k_2\rangle \cong G$, and $G \in \text{Sk}\,M$.

If G has solvable word problem, then (2) holds in every existentially closed group. If not, there are existentially closed groups in which (2) holds, and others in which it does not hold.

We note that we do not really need G to be finitely presented. As long as we can find a finitely presented group $H = \langle g_1, g_2 \mid R(g_1, g_2)\rangle$ containing G, so that

$$G = \langle p(g_1, g_2), \quad q(g_1, g_2)\rangle$$

(say), we can replace

$$\mathrm{Hom}\,(x_1, x_2, y_1, y_2, s, t)$$

by

$$\mathrm{Hom}\,(p(x_1, x_2),\quad q(x_1, x_2),\quad p(y_1, y_2),\quad q(y_1, y_2),\quad s, t),$$

and get the same result.

To see this, suppose that

$$G \in \mathrm{Sk}\, M,$$

and let $\bar{p}, \bar{q} \in M$ generate a group isomorphic to G. Then the equations

$$R(x_1, x_2) \cup \{q(x_1, x_2) = q(g_1, g_2),\quad p(x_1, x_2) = p(g_1, g_2)\}$$

can be solved in the amalgamated free product $M *_G H$. So we can choose $k_1, k_2 \in M$ such that $R(k_1, k_2)$ is true in M, and such that $\langle p(\boldsymbol{k}), q(\boldsymbol{k})\rangle = G$. Then if $h_1, h_2 \in M$, and if $R(h_1, h_2)$ is true in M, we have that $\langle p(\boldsymbol{h}), q(\boldsymbol{h})\rangle$ is a homomorphic image of G, and so the formula

$$(\exists x_1,x_2)(\forall y_1,y_2)(\exists s,t)\Big(R(x_1, x_2)\ \&$$
$$\Big(R(y_1, y_2) \Rightarrow \mathrm{Hom}\,(p(x_1, x_2), q(x_1, x_2), p(y_1, y_2), q(y_1, y_2), s, t)\Big)\Big) \quad (2')$$

is true in M.

Similarly, if we can choose solutions k_1 and k_2 in M, and if there is some relation $w(p, q) = 1$ that holds in M but is not a consequence of $R(x_1, x_2)$, then there exist $h_1, h_2 \in M$ such that there is no homomorphism carrying $p(k_1, k_2)$ to $p(h_1, h_2)$ and $q(k_1, k_2)$ to $q(h_1, h_2)$. Thus (2′) holds for M if and only if $G \in \mathrm{Sk}\, M$.

For any two groups G and H, we can define $\mathbf{N}_{G,H}$ to be the set of all kernels of all homomorphisms from G to H. So

$$\mathbf{N}_{G,H} = \{N \mid N \trianglelefteq G,\quad G/N \quad \text{is embedded in } H\},$$

and we suppose that $\mathbf{N}_{G,H}$ is partially ordered under inclusion. If G is a finitely presented group, and if M is an existentially closed group, then any first-order property of $\mathbf{N}_{G,M}$, as a partially ordered set, can be translated into a first-order sentence that is true in M if and only if the property is true for $\mathbf{N}_{G,M}$. We translate properties of N into properties of the homomorphic image G/N of G in M. For example, if $G = \langle x, y \mid R(x, y)\rangle$, if $N_1, N_2 \in \mathbf{N}_{G,M}$, and if $\langle x_i, y_i\rangle$ is the image of G/N_i $(1 \leqslant i \leqslant 2)$ in M, then $N_1 \leqslant N_2$ if and only if the sentence $(\exists s,t)\,\mathrm{Hom}\,(x_1, y_1, x_2, y_2, s, t)$ is true in M. So the property '$N_1 \leqslant N_2$' has become 'G/N_1 is embedded in G/N_2'. The statement '$\exists N$' translates to the statement $(\exists x, y)R(x, y)$, (so $\langle x, y\rangle = G/N$, for some N), and properties of the form $(\exists N)(\cdots)$ and $(\forall N)(\cdots)$ become sentences of the form

$$(\exists x,y)\big(R(x, y) \,\&\, (\cdots)\big)$$

and

$$(\forall x, y)\big(R(x, y) \Rightarrow (\cdots)\big),$$

respectively. The sentence (2) is equivalent to the statement $(\exists N_1)(\forall N_2)(N_1 \leqslant N_2)$, i.e. (2) is true in an existentially closed group M if and only if $\mathbf{N}_{G,M}$ has a minimal element.

Since $G \in \mathbf{N}_{G,M}$, it follows that $\mathbf{N}_{G,M}$ always has a maximal element. The next example is a first order-sentence that is true in an existentially closed group M if and only if the set $\mathbf{N}_{G,M}\backslash\{G\}$ has a maximal element.

Example 3 Let $\langle g_1, \ldots, g_n \mid R(\boldsymbol{g}) \rangle = G$ be a finitely presented group. Consider the sentence

$$\begin{aligned}&(\exists \boldsymbol{x})\Big(R(\boldsymbol{x}) \,\&\, (x_1 \neq 1 \text{ or} \cdots \text{or } x_n \neq 1) \,\&\\ &\quad (\forall \boldsymbol{y})\big((\exists s,t) \operatorname{Hom}(x_1, \ldots, x_n, y_1, \ldots, y_n, s, t) \Rightarrow\\ &\quad ((y_1 = 1 \,\&\cdots\&\, y_n = 1) \text{ or } (\exists w)(w^{-1}y_1w = x_1 \,\&\cdots\&\, w^{-1}y_nw = x_n))\big)\Big).\end{aligned} \tag{3}$$

This holds in M if and only if M contains a non-trivial homomorphic image $\langle \bar{x}_1, \ldots, \bar{x}_n\rangle$ of G that has no proper homomorphic image in M.

Thus, if (3) is true in M, and if θ is a homomorphism which carries g_i to $\bar{x}_i$ $(1 \leqslant i \leqslant n)$, then $\ker\theta \in \mathbf{N}_{G,M}\backslash\{G\}$, and, if $\ker\theta \subseteq N$, for some $N \in \mathbf{N}_{G,M}$, then G/N is a homomorphic image of $G/\ker\theta$. So $N = G$ or $N = \ker\theta$. Thus, $\ker\theta$ is a maximal element of $\mathbf{N}_{G,M}\backslash\{G\}$. If N is a maximal element of $\mathbf{N}_{G,M}\backslash\{G\}$, then, provided that

$$(G/N) \cong \langle \bar{x}_1, \ldots, \bar{x}_n\rangle \subseteq M,$$

we have that $\bar{x}_1, \ldots, \bar{x}_n$ are solutions, in M, for (3). So (3) is true in M if and only if $\mathbf{N}_{G,M}\backslash\{G\}$ has a maximal element.

Miller (1973) has an example of a finitely presented group G, every non-trivial homomorphic image of which has unsolvable word problem. For this G, (3) is false in $\mathrm{M}_\varnothing$, because every finitely generated simple subgroup of $\mathrm{M}_\varnothing$ has a recursively enumerable relation set, and hence has solvable word problem. So $\mathrm{M}_\varnothing$ cannot contain a non-trivial homomorphic image of G. But every finitely generated group has a simple homomorphic image, and G is not simple, so G has a simple proper homomorphic image G/N, and this image is embeddable in some existentially closed group M. Since G/N is simple, it has no proper homomorphic images at all. So (3) is true in this M, and there exist existentially closed groups in which (3) is true when G is taken to be Miller's group.

9.4 ARITHMETICALLY RELATED GROUPS

In this section we consider types of finitely generated groups that have an associated first-order sentence specifying them in existentially closed groups.

Definition 9.7 The set $\mathfrak{A}$ of *arithmetic sets* is the smallest set of subsets of $\mathbb{N}$ that contains the empty set and that is closed under enumeration reducibility and the taking of complements; i.e. if $X \in \mathfrak{A}$ then $(\mathbb{N}\backslash X) \in \mathfrak{A}$, and if $Y \leqslant_e X$ then $Y \in \mathfrak{A}$. A set is called *arithmetic* if it lies in $\mathfrak{A}$.

[The usual logical definition says that an arithmetic set is a set of the form

$$\{x_1 \mid (Q_2x_2)\cdots(Q_nx_n)((x_1, \ldots, x_n) \in R)\}, \qquad (*)$$

where $R \subseteq \mathbb{N}^n$ is a recursive set, and $Q_i \in \{\exists, \forall\}$ $(2 \leqslant i \leqslant n)$. These two definitions can be seen to be equivalent, as follows.

Let $\mathfrak{B}$ be the set of all sets of the form (*). Suppose that $X \in \mathfrak{B}$ has the form (*), and proceed by induction on n.

If $n = 1$, then $X = R$, so X is recursive and $X \in \mathfrak{A}$ (since $X \leqslant_e \varnothing$).

We now assume that $n \geqslant 2$, and that the result is true for $m \leqslant n-1$. Suppose first that $Q_2 = \exists$, and let

$$Y = \{(x_1, x_2) \mid (Q_3x_3)\cdots(Q_nx_n)((x_1, \ldots, x_n) \in R)\}.$$

Then $x_1 \in X$ if and only if, for some x_2, $(x_1, x_2) \in Y$. Thus

$$X \leqslant_e Y \quad \text{via} \quad U = \{(x_1, \{x_1, x_2\}) \mid (x_1, x_2, \ldots, x_n) \in R\}.$$

Let $f: \mathbb{N} \times \mathbb{N} \to \mathbb{N}$ be a recursive bijection, and let

$$Z = \{f(x_1, x_2) \mid (x_1, x_2) \in Y\},$$
$$S = \{(f(x_1, x_2), x_3, \ldots, x_n) \mid (x_1, \ldots, x_n) \in R\}.$$

Then $Y \equiv_1 Z$, $S \equiv_1 R$, and

$$Z = \{x \mid (Q_3x_3)\cdots(Q_nx_n)((x, x_2, \ldots, x_n) \in S)\}.$$

Since S is recursive, we have that $Z \in \mathfrak{B}$, and so, by induction, $Z \in \mathfrak{A}$. But $X \leqslant_e Z$, so $X \in \mathfrak{A}$.

If $Q_2 = \forall$, then

$$X = \mathbb{N}\backslash X_1,$$

where

$$X_1 = \{x_1 \mid (\exists x_2)(\bar{Q}_3x_3)\cdots(\bar{Q}_nx_n)((x_1, \ldots, x_n) \in \mathbb{N}^n\backslash R)\},$$

(here: $\{\bar{Q}_i\} = \{\exists, \forall\}\backslash\{Q_i\}$). If R is recursive, then so is $\mathbb{N}^n\backslash R$; so, by the first case, we have $X_1 \in \mathfrak{A}$. But $\mathfrak{A}$ is closed under taking complements, so

$X \in \mathfrak{A}$. Thus, by induction,

$$\mathfrak{B} \subseteq \mathfrak{A}.$$

Clearly, $\varnothing \in \mathfrak{B}$, and $\mathfrak{B}$ is closed under taking complements. So, to show that $\mathfrak{A} \subseteq \mathfrak{B}$, we show that $\mathfrak{B}$ is closed under enumeration reducibility.

Suppose that $X \in \mathfrak{B}$ has the form (*), and suppose that $X \leqslant_e Y$ via U. Enumerate $\mathscr{P}_f(\mathbb{N})$ as $D_0, D_1, D_2, \ldots$, and let

$$V = \{(p, m) \mid (p, D_m) \in U\}.$$

Then we have that

$$Y = \{p \mid (\exists m)((p, m) \in V \quad \& \quad D_M \subseteq X)\}.$$

Now, $D_M \subseteq X$ if and only if, for each $y \in \mathbb{N}$, either $y \in X$ or $y \notin D_M$. So the set of all m such that $D_M \subseteq X$ is the set

$$\{m \mid (\forall y)(Q_2 y_2) \cdots (Q_n y_n)((y, y_2, \ldots, y_n) \in R \quad \text{or} \quad y \in \mathbb{N} \backslash D_m)\}.$$

Let S and T be the sets

$$\{(p, m, y, y_2, \ldots, y_n) \mid (p, m) \in V, \quad y \in \mathbb{N} \backslash D_m, \quad (y_2, \ldots, y_n) \in \mathbb{N}^{n-1}\},$$
$$\{(p, m, y, y_2, \ldots, y_n) \mid (p, m) \in V, \quad (y, y_2, \ldots, y_n) \in R\},$$

respectively. Then $p \in Y$ if and only if, for some m, we have

$$(p, m) \in V \text{ and } D_m \subseteq X,$$

that is, if and only if there is an m such that $(p, m) \in V$ and, for all $y \in \mathbb{N}$, either $y \in \mathbb{N} \backslash D_m$ or

$$(Q_2 y_2) \cdots (Q_n y_n)((y, y_2, \ldots, y_n) \in R).$$

So we have

$$Y = \{p \mid (\exists m)(\forall y)(Q_2 y_2) \cdots (Q_n y_n)((p, m, y, y_2, \ldots, y_n) \in S \cup T)\}.$$

Now, $S \cup T$ is recursively enumerable; so, by a well-known theorem (see Rogers, 1967, p. 66), there is a recursive set R_1 such that

$$S \cup T = \{(x_1, \ldots, x_{n+2}) \mid (\exists x_{n+3})((x_1, \ldots, x_{n+3}) \in R_1)\}.$$

So,

$$Y = \{x_1 \mid (\exists x_2)(\forall x_3)(Q_4 x_4) \cdots (Q_{n+2} x_{n+2})(\exists x_{n+3})((x_1, x_2, \ldots, x_{n+3}) \in R_1)\},$$

and $Y \in \mathfrak{B}$, as required.]

We recall that, if $Y \leqslant_e X$, then $Y \leqslant_e X$ via U, for some recursively enumerable subset U of $\mathbb{N} \times \mathscr{P}_f(\mathbb{N})$, and that the set of all such recursively enumerable subsets can be listed, say $U_0, U_1, U_2, \ldots$, in such a way that the set $\{(m, n, A) \mid (n, A) \in U_m\}$ is recursively enumerable (see Chapter 4).

Let

$$U_i(X) = \{n \mid (n, A) \in U_i, \quad A \subseteq X\},$$

so that $U_i(X)$ is the unique subset of $\mathbb{N}$ that is enumeration-reducible to X via U_i. Define a map

$$\varepsilon_i : \mathscr{P}(\mathbb{N}) \to \mathscr{P}(\mathbb{N})$$

by the rule $X\varepsilon_i = U_i(X)$. Let

$$\kappa : \mathscr{P}(\mathbb{N}) \to \mathscr{P}(\mathbb{N})$$

be the map defined by the rule $X\kappa = \mathbb{N} \backslash X$, for all $X \subseteq \mathbb{N}$. The set $\{\kappa, \varepsilon_i \mid i \in \mathbb{N}\}$ generates a semigroup Σ, under composition of maps. Since Σ is countably generated, Σ is countable, and, since $\mathfrak{A} = \{\varnothing\tau \mid \tau \in \Sigma\}$, where $\varnothing$ is the empty set, we have that $\mathfrak{A}$ is countable. Every element of Σ can be written in the form

$$\varepsilon_{i_1}\kappa\varepsilon_{i_2}\kappa \cdots \varepsilon_{i_n} \tag{$*$}$$

for some $n \in \mathbb{N}$. If $X = \varnothing\tau$, and if τ contains n occurrences of κ when written in the form $(*)$, we shall say that X is arithmetic at *level n*. Note: we may have $U_i(X) = U_j(X)$ with $i \neq j$, so there exist τ_1 and τ_2 such that $\varnothing\tau_1 = \varnothing\tau_2$, but $\tau_1 \neq \tau_2$. Thus X may be arithmetic at several different levels. In fact, the set $U = \{(n, \{n\}) \mid n \in \mathbb{N}\}$ is one of the U_j, for some j. If $U = U_j$ then $U(X) = X$ and ε_j is the identity, so $\varnothing\tau = \varnothing\tau\kappa\varepsilon_j\kappa$ and every set which is arithmetic at level n is also arithmetic at level $n+2$.

If G is a finitely presented group whose relations form an arithmetic set, then, within first-order group theory, there exists a sentence that is true in an existentially closed group M if and only if $G \in \operatorname{Sk} M$. The next sequence of lemmas and theorems leads to a proof of this result.

Recall, from Chapter 4, the group G_X which is generated by the standard generators a and b, subject to the defining relations

$$\{ac_i = c_i a, \quad bc_i = c_i b, \quad c_i^2 = 1, \quad c_j = 1 \mid i \in \mathbb{N}, \quad j \in X\}$$
$$(\text{where } b_i = a^{-i}ba^i \quad \text{and} \quad c_i = [b, b_{i+1}]),$$

with the property that $\operatorname{Rel} \mathrm{G}_X \equiv^* X$. If X is recursive, then G_X has solvable word problem, and so $\mathrm{G}_X \in \operatorname{Sk} M$, for every existentially closed group M. Clearly, there is a homomorphism $\mathrm{G}_X \to \mathrm{G}_Y$ that maps the standard generators to the standard generators, if and only if $X \subseteq Y$.

For any $X \subseteq \mathbb{N}$, the set X is equal to the union of all the recursive sets that it contains, and is also equal to the intersection of all the recursive sets containing it. For example, every 1-element set is recursive so, clearly,

$$X=\bigcup\{\{x\} \mid x\in X\} \subseteq \bigcup\{Z \mid Z \text{ recursive}, \quad Z\subseteq X\} \subseteq X,$$

$$\begin{aligned} X &= \mathbb{N}\backslash(\mathbb{N}\backslash X)=\mathbb{N}\backslash(\bigcup\{Z \mid Z \text{ recursive}, \quad Z\subseteq \mathbb{N}\backslash X\}) \\ &=\bigcap\{\mathbb{N}\backslash Z \mid Z \text{ recursive}, \quad Z\cap X=\varnothing\} \\ &=\bigcap\{Y \mid Y \text{ recursive}, \quad X\subseteq Y\}. \end{aligned}$$

Thus, in some sense, any group G_X can be 'approximated' above and below by groups with solvable word problem. This approximation helps us to prove the result.

For any set $\Delta\subseteq\mathscr{P}(\mathbb{N})$, let $St(x, y; \Delta)$ be a formula of L with variables x and y, such that, for any existentially closed group M,

$$St(x, y; \Delta) \quad \text{is true at} \quad (M, \bar{x}, \bar{y})$$

if and only if

$$\bar{x} \text{ and } \bar{y} \text{ are standard generators for } G_X, \text{ for some } X\in\Delta.$$

For any given Δ, a formula with this property may not exist, but if there do exist formulae with this property, we pick one, arbitrarily, and call it $St(x, y; \Delta)$. For ease of notation, we will write $St(x, y; X)$ for $St(x, y; \{X\})$, and we will write $St(x, y)$ for $St(x, y; \mathscr{P}(\mathbb{N}))$.

We let $\bar{\mathfrak{A}}$ denote the set of all subsets $X\subseteq\mathbb{N}$ for which a formula $St(x, y; X)$ exists. Our first aim is to show that $\mathfrak{A}\subseteq\bar{\mathfrak{A}}$.

To begin with, we want to show that there exists a formula $St(x, y)$. It is fairly easy to produce a recursively enumerable set $\mathscr{S}$ of equations, inequalities and positive implications, defined on x and y over the trivial group, such that $\bar{x}$ and $\bar{y}$ are solutions for $\mathscr{S}$ if and only if they are standard generators of some G_X. The idea is to replace $\mathscr{S}$ by a finite set $\mathscr{S}^*$ of equations and inequalities on the variables

$$x, y, x_1, \ldots, x_n,$$

such that $\bar{x}$ and $\bar{y}$ are solutions for $\mathscr{S}$ if and only if, for some $\bar{x}_1, \ldots, \bar{x}_n$, the elements $\bar{x}, \bar{y}, \bar{x}_1, \ldots, \bar{x}_n$ are solutions for $\mathscr{S}^*$. For then we see that, for any existentially closed group M, and for any $\bar{x}, \bar{y}\in M$, the formula

$$(\exists x_1, \ldots, x_n)P(x, y, x_1, \ldots, x_n),$$

where P is the conjunction of the equations and inequalities in $\mathscr{S}^*$, is true at $(M, \bar{x}, \bar{y})$ if and only if $\bar{x}$ and $\bar{y}$ satisfy $\mathscr{S}$. This is the sort of thing that Theorem 5.3 allows us to do; but Theorem 5.3 is not quite what is required, since our set $\mathscr{S}$ will be defined over the trivial group. Of course, for each existentially closed group M, the set $\mathscr{S}$ is defined over some non-trivial subgroup of M, but the set $\mathscr{S}^*$ in Theorem 5.3 depends on this non-trivial group non-trivially when $\mathscr{S}$ contains inequalities; i.e. $\mathscr{S}^*$ is not defined over the trivial group, but over $\langle g \rangle$, where g is some element of $M\backslash\{1\}$. So the conjunction $P(x, y, x_1, \ldots, x_n)$ would involve group elements and would not be a formula in L. Thus, we consider the following variation of Theorem 5.3.

Theorem 9.8 *Let $\mathcal{S}(x_1, \ldots, x_n)$ be any recursively enumerable set of equations, inequalities, and implications defined over some finitely generated group $G = \langle g_1, \ldots, g_r \rangle$. There exists a finite set $\mathcal{S}^*(x_1, \ldots, x_{n+l})$, defined over G and consisting of equations and one inequality, such that $\bar{x}_1, \ldots, \bar{x}_n$ are solutions for $\mathcal{S}$ if and only if, for some $\bar{x}_{n+1}, \ldots, \bar{x}_{n+l}$, the elements $\bar{x}_1, \ldots, \bar{x}_{n+l}$ are solutions for $\mathcal{S}^*$.*

Proof Choose a new variable x_{n+1}. We obtain a set $\mathcal{S}_1(x_1, \ldots, x_{n+1})$ from $\mathcal{S}$ by adding the inequality $x_{n+1} \neq 1$, replacing each inequality $\omega \neq 1$ in $\mathcal{S}$ by a corresponding implication

$$(w = 1) \Rightarrow (x_{n+1} = 1),$$

and replacing each negative implication

$$(v_1 = 1) \,\&\cdots\&\, (v_k = 1) \Rightarrow \Omega$$

in $\mathcal{S}$, by a positive implication

$$(v_1 = 1) \,\&\cdots\&\, (v_k = 1) \Rightarrow (x_{n+1} = 1).$$

It is easy to see that $\bar{x}_1, \ldots, \bar{x}_n$ are solutions for $\mathcal{S}$ if and only if there exists an element $\bar{x}_{n+1}$ such that $\bar{x}_1, \ldots, \bar{x}_{n+1}$ are solutions for $\mathcal{S}_1$.

As in the proof of Theroem 5.3, we obtain a set

$$\mathcal{S}_2(x_1, \ldots, x_{n+1}, z_0, z_1, \ldots)$$

of equations, together with the inequality $x_{n+1} \neq 1$, from $\mathcal{S}_1$ by replacing each implication of the form

$$(v_1 = 1) \,\&\cdots\&\, (v_k = 1) \Rightarrow (u = 1)$$

by an equation of the form

$$z_{i_1}^{-1} v_1 z_{i_1} z_{j_1}^{-1} v_1 z_{j_1} \cdots z_{i_k}^{-1} v_k z_{i_k} z_{j_k}^{-1} v_k z_{j_k} u^{-1} = 1.$$

From Lemma 1.6, we see that $\bar{x}_1, \ldots, \bar{x}_{n+1}$ are solutions for $\mathcal{S}_1$ if and only if there exists elements $\bar{z}_0, \bar{z}_1, \ldots$ such that $\bar{x}_1, \ldots, \bar{x}_{n+1}, \bar{z}_0, \bar{z}_1, \ldots$ are solutions for $\mathcal{S}_2$.

Let

$$\{e_i(x_{n+2}, x_{n+3}) \mid i \in \mathbb{N}\}$$

be the recursive set defined in Lemma 5.1. we obtain

$$\mathcal{S}_3(x_1, \ldots, x_{n+1}, x_{n+2}, x_{n+3})$$

from $\mathcal{S}_2$ by substituting $\mathrm{e}_i(x_{n+2}, x_{n+3})$ for each occurrence of z_i $(i \in \mathbb{N})$ in

$\mathcal{S}_2$. (Note, in Theorem 5.3 we substituted some e_i for every variable in $\mathcal{S}_2$. But here we want the original variables $x_1, \ldots, x_n$ also to appear in the final set $\mathcal{S}^*$, so we do not substitute for them. This is why we need the assumption that $\mathcal{S}$ is defined over finitely many variables.) Then $\bar{x}_1, \ldots, \bar{x}_{n+3}$ are solutions for $\mathcal{S}_3$ if and only if there exist elements $\bar{e}_0$, $\bar{e}_1$, $\bar{e}_2, \ldots$ such that

$$\bar{x}_1, \ldots, \bar{x}_{n+1}, \bar{e}_0, \bar{e}_1, \bar{e}_2, \ldots$$

are solutions for $\mathcal{S}_2$, and in this case we can take

$$\bar{e}_i = \mathrm{e}_i(\bar{x}_{n+2}, \bar{x}_{n+3}) \quad (i \in \mathbb{N}).$$

So, there exist solutions $\bar{x}_1, \ldots, \bar{x}_{n+3}$ for $\mathcal{S}_3$ if and only if $\bar{x}_1, \ldots, \bar{x}_n$ are solutions for $\mathcal{S}$.

The group

$$H = \langle h_1, \ldots, h_{r+n+3} \mid w(\boldsymbol{h}) = 1, \quad \text{for all} \quad w(\boldsymbol{g}, \boldsymbol{x}) = 1 \quad \text{in } \mathcal{S}_3 \rangle$$

is a subgroup of some finitely presented group

$$K = \langle k_1, \ldots, k_m \mid r_1(\boldsymbol{k}) = \cdots = r_s(\boldsymbol{k}) = 1 \rangle,$$

and, if $\bar{x}_1, \ldots, \bar{x}_{n+3}$ are solutions to $\mathcal{S}_3$, then there is a homomorphism $\theta : H \to \langle G, \bar{x}_1, \ldots, \bar{x}_{n+3} \mid \mathcal{S}_3 \rangle$ that satisfies

$$h_j\theta = g_i \quad (1 \leqslant j \leqslant r), \qquad h_{r+k}\theta = \bar{x}_k \quad (1 \leqslant k \leqslant n+3).$$

Choose new variables $y_1, \ldots, y_m, y, z$, and words

$$v_i(y_1, \ldots, y_m) \qquad\qquad (1 \leqslant i \leqslant r+n+3)$$

such that $v_i(\boldsymbol{k}) = h_i$. (From the statement of the theorem, we really want to call these new variables $x_{n+4}, \ldots, x_{n+m+5}$ but it is easier, notationally and conceptually, to use different types of names for sets of variables that do different types of job: it is easier to see what is going on.)

Define $\mathcal{S}^*(x_1, \ldots, x_{n+3}, y_1, \ldots, y_m, y, z)$ to be the set

$$\left\{ \begin{array}{c|c} r_p(y_1, \ldots, y_m) = 1 & 1 \leqslant p \leqslant s \\ [v_i, y^{-1}g_j y] = 1 & 1 \leqslant i \leqslant r+n+3 \\ [v_i, y^{-1}x_k y] = 1 & 1 \leqslant j \leqslant r \\ v_j^{-1}y^{-1}g_j y = z^{-1}v_j z & 1 \leqslant k \leqslant n+3 \\ v_{r+k}y^{-1}x_k y = z^{-1}v_{r+k}z & \\ x_{n+1} \neq 1 & \end{array} \right\}$$

If $\bar{x}_1, \ldots, \bar{x}_n$ are solutions to $\mathcal{S}$, we want to show that we can find elements $\bar{x}_{n+1}, \bar{x}_{n+2}, \bar{x}_{n+3}, \bar{y}_1, \ldots, \bar{y}_m, \bar{y}, \bar{z}$ such that $\bar{x}_1, \ldots, \bar{x}_{n+3}$, $\bar{y}_1, \ldots, \bar{y}_m$, $\bar{y}, \bar{z}$ are solutions to $\mathcal{S}^*$. We know that we can find $\bar{x}_{n+1}$, $\bar{x}_{n+2}$, $\bar{x}_{n+3}$ such that $\bar{x}_1, \ldots, \bar{x}_{n+3}$ are solutions for $\mathcal{S}_3$. Take

$$\bar{y}_l = k_l \;\; (1 \leqslant l \leqslant m)$$

so that $\bar{x}_1, \ldots, \bar{x}_{n+3}, \bar{y}_1, \ldots, \bar{y}_m$ are solutions, in $\langle G, \bar{x}_1, \ldots, \bar{x}_{n+3}\rangle \times K$, for $\mathcal{S}_3 \cup \{r_p = 1 \mid 1 \leqslant p \leqslant s\}$. Since $x_{n+1} \neq 1$ lies in $\mathcal{S}_3$, we have $\bar{x}_{n+1} \neq 1$. Let

$$\bar{v}_i = v_i(\bar{y}_1, \ldots, \bar{y}_m) = h_i \quad (1 \leqslant i \leqslant n + r + 3).$$

Then, as above, $\langle g_1, \ldots, g_r, \bar{x}_1, \ldots, \bar{x}_{n+3}\rangle$ is a homomorphic image of $\langle \bar{v}_1, \ldots, \bar{v}_{r+n+3}\rangle$. So, by Lemma 5.2, there exist elements $\bar{y}$ and $\bar{z}$ satisfying $\mathcal{S}^*(\bar{x}_1, \ldots, \bar{x}_{n+3}, \bar{y}_1, \ldots, \bar{y}_m, y, z)$. Thus $\bar{x}_1, \ldots, \bar{x}_{n+3}, \bar{y}_1, \ldots, \bar{y}_m, \bar{y}, \bar{z}$ are solutions for $\mathcal{S}^*$.

Conversly, suppose that $\bar{x}_1, \ldots, \bar{x}_{n+3}, \bar{y}_1, \ldots, \bar{y}_m, \bar{y}, \bar{z}$ are solutions for $\mathcal{S}^*$. To see that $\bar{x}_1, \ldots, \bar{x}_n$ are solutions for $\mathcal{S}$, it is sufficient to show that $\bar{x}_1, \ldots, \bar{x}_{n+3}$ are solutions for $\mathcal{S}_3$.

Since $x_{n+1} \neq 1$ belongs to $\mathcal{S}^*$, we have $\bar{x}_{n+1} \neq 1$.

Suppose that $w(g_1, \ldots, g_r, x_1, \ldots, x_{n+3}) = 1$ is an equation in $\mathcal{S}_3$. Then $w(\boldsymbol{h}) = 1$, and, since $r_p(\bar{y}_1, \ldots, \bar{y}_m) = 1$ $(1 \leqslant p \leqslant s)$, we have that $w(\bar{v}_1, \ldots, \bar{v}_{n+r+3}) = 1$, where $\bar{v}_i = v_i(\bar{y}_1, \ldots, \bar{y}_m)$. Since $\bar{y}$ and $\bar{z}$ are solutions for

$$\mathcal{S}^*(\bar{x}_1, \ldots, \bar{x}_{n+3}, \bar{y}_1, \ldots, \bar{y}_m, y, z),$$

it follows, by Lemma 5.2, that there is a homomorphism

$$\theta : \langle \bar{v}_1, \ldots, \bar{v}_{n+r+3}\rangle \to \langle g_1, \ldots, g_r, \bar{x}_1, \ldots, \bar{x}_{n+3}\rangle.$$

So $w(g_1, \ldots, g_r, \bar{x}_1, \ldots, \bar{x}_{n+3}) = 1$. Thus $\bar{x}_1, \ldots, \bar{x}_{n+3}$ are solutions for $\mathcal{S}_3$.

Re-naming $y_1 = x_{n+4}, \ldots, y_m = x_{n+m+3}$, $y = x_{n+m+4}$, $z = x_{n+m+5}$, this proves the theorem. □

Lemma 9.9 *There exists a formula $St(x, y)$.*

Proof From the definition of G_X, it is easy to see that $\bar{x}$ and $\bar{y}$ are standard generators for some G_X if and only if they are solutions for the set

$$\begin{aligned}\mathcal{S} = \{&w(x, y) = 1 \mid w \in \mathrm{Rel}\, G_\varnothing\} \\ &\cup \{w(x, y) \neq 1 \mid w \in \mathrm{Nonrel}\, G_\mathbb{N}\} \\ &\cup \{(c_{i_1} \cdots c_{i_n} = 1) \Rightarrow (c_{i_1} = 1) \mid \{i_1, \ldots, i_n\} \subseteq \mathbb{N},\quad n \in \mathbb{N}\},\end{aligned}$$

where $c_i = [y, x^{-i-1} y x^{i+1}]$. By Theorem 9.8, we can find a finite set

$$\mathcal{S}^* = \{u_i(x, y, x_1, \ldots, x_n) = 1,\quad v_j(x, y, x_1, \ldots, x_n) \neq 1 \mid 1 \leqslant i \leqslant m,\quad 1 \leqslant j \leqslant k\}$$

such that $\bar{x}$ and $\bar{y}$ are solutions to $\mathcal{S}$ if and only if $\mathcal{S}^*(\bar{x}, \bar{y}, x_1, \ldots, x_n)$ has a solution. So, we may take $St(x, y)$ to be the formula

$$(\exists x_1, \ldots, x_n)((u_1 = 1) \,\&\, \cdots \,\&\, (u_m = 1) \,\&\, (v_1 \neq 1) \,\&\, \cdots \,\&\, (v_k \neq 1)).$$ □

We now use Lemma 9.9 to prove the next lemma, which, in turn, allows us to prove that $\mathfrak{A} \subseteq \bar{\mathfrak{A}}$.

Lemma 9.10 *For any $X \subseteq \mathbb{N}$, if any one of the following four formulae exists, then they all do*:

(i) $St(x, y; \{Y \mid X \subseteq Y\})$,
(ii) $St(x, y; \{Y \mid Y \subseteq X\})$,
(iii) $St(x, y; \{Y \mid X \cap Y = \varnothing\})$,
(iv) $St(x, y; \{Y \mid X \cup Y = \mathbb{N}\})$.

Proof For fixed $X \subseteq \mathbb{N}$ let

$$\Delta_1 = \{Y \mid X \subseteq Y\}, \qquad \Delta_3 = \{Y \mid X \cap Y = \varnothing\},$$
$$\Delta_2 = \{Y \mid Y \subseteq X\}, \qquad \Delta_4 = \{Y \mid X \cup Y = \mathbb{N}\}.$$

We will show that the existence of (i) is equivalent to the existence of (ii), then that (iii) being equivalent to (iv) is a corollary of this, and finally that (i) is equivalent to (iv).

Let $C(x, y, u, v)$ denote the formula

$$St(x, y) \,\&\, St(u, v) \,\&\, (\exists r, s) \operatorname{Hom}(x, y, u, v, r, s).$$

Then $C(x, y, u, v)$ is true at $(M, \bar{x}, \bar{y}, \bar{u}, \bar{v})$ if and only if $(\bar{x}, \bar{y})$ and $(\bar{u}, \bar{v})$ are pairs of standard generator for some G_Y and G_Z, respectively, with $Y \subseteq Z$.

Suppose that (i) exists, and consider the formula

$$St(x, y) \,\&\, (\forall u, v)\big(St(u, v; \Delta_1) \Rightarrow C(x, y, u, v)\big). \tag{1}$$

We show that (1) is a formula of the form $St(x, y; \Delta_2)$; so we want to show that (1) is true at $(M, \bar{x}, \bar{y})$ if and only if $\bar{x}$ and $\bar{y}$ are the standard generators of G_Y, for some $Y \subseteq X$.

Given any recursive set Z and any existentially closed group M, then $G_Z \in \operatorname{Sk} M$. Thus, for each recursive set Z, we can find standard generators $\bar{u}, \bar{v} \in M$ for G_Z, and if $X \subseteq Z$ then $St(u, v; \Delta_1)$ will be true at $(M, \bar{u}, \bar{v})$. So, if (1) is true at $(M, \bar{x}, \bar{y})$, then, for each recursive set $Z \supseteq X$, there exist $\bar{u}, \bar{v} \in M$, that are standard generators for G_Z, such that $C(x, y, u, v)$ is true at $(M, \bar{x}, \bar{y}, \bar{u}, \bar{v})$. So $\langle \bar{x}, \bar{y}\rangle \cong G_Y$, for some $Y \subseteq Z$. Since $X = \bigcap\{Z \mid Z \text{ recursive}, \ X \subseteq Z\}$, we have $Y \subseteq X$, as required. If $\bar{x}, \bar{y} \in M$ are standard generators for G_Y, for some $Y \subseteq X$, then $St(x, y)$ is true in M. Clearly, if $St(u, v; \Delta_1)$ is true at $(M, \bar{u}, \bar{v})$, then $\langle \bar{u}, \bar{v}\rangle \cong G_Z$, for some $Z \supseteq X \supseteq Y$, and $C(\bar{x}, \bar{y}, \bar{u}, \bar{v})$ is true in M, and thus (1) is true at $(M, \bar{x}, \bar{y})$. Hence

$$St(x, y) \,\&\, (\forall u, v)\big(St(u, v; \Delta_1) \Rightarrow C(x, y, u, v)\big)$$

is true at $(M, \bar{x}, \bar{y})$ if and only if $\bar{x}$ and $\bar{y}$ are standard generators for G_Y, for some $Y \supseteq X$. So, this formula is of the form $St(x, y; \Delta_2)$, and we have shown that (i)$\Rightarrow$(ii).

Since it is also true that $X = \bigcup\{Z \supseteq X \mid Z \text{ recursive}\}$, we see that

$$St(x, y) \,\&\, (\forall u, v)\big(St(u, v; \Delta_2) \Rightarrow C(u, v, x, y)\big)$$

is of the form $St(x, y; \Delta_1)$, so that (ii)$\Rightarrow$(i).

There is a formula of the form $St(x, y; \Delta_4)$ if and only if there is a formula $St\big(x, y; \{Y \mid (\mathbb{N}\backslash X) \subseteq Y\}\big)$. Since (i)$\Leftrightarrow$(ii), this is equivalent to the existence of a formula $St(x, y; \{Y \mid Y \subseteq \mathbb{N}\backslash X\})$, and hence to the existence of $St(x, y; \Delta_3)$. So (iii)$\Leftrightarrow$(iv).

By Lemma 9.9, a formula of the form $St(x, y; \{Y \mid Y \subseteq \mathbb{N}\})$ exists. So, since (i)$\Leftrightarrow$(ii), a formula of the form

$$St(x, y; \{Y \mid \mathbb{N} \subseteq Y\}) = St(x, y; \mathbb{N}),$$

exists.

We let $U(x, y, u, v)$ denote the formula

$$(\forall p, q)\Big(\big(C(x, y, p, q) \,\&\, C(u, v, p, q)\big) \Rightarrow St(p, q; \mathbb{N})\Big).$$

Then $U(x, y, u, v)$ is true at $(M, \bar{x}, \bar{y}, \bar{u}, \bar{v})$ if and only if, whenever $(\bar{x}, \bar{y})$ and $(\bar{u}, \bar{v})$ are standard generator pairs for G_X and G_Y, then $X, Y \subseteq Z$ implies that $Z = \mathbb{N}$. In other words, $X \cup Y = \mathbb{N}$.

Consider the formula

$$St(x, y) \,\&\, (\forall u, v)\big(St(u, v; \Delta_1) \Rightarrow U(x, y, u, v)\big). \tag{2}$$

If (2) is true at $(M, \bar{x}, \bar{y})$, where $\bar{x}$ and $\bar{y}$ are standard generators for G_Z, then, for all recursive sets $Y \supseteq X$, there exist $\bar{u}, \bar{v} \in M$ that are standard generators for G_Y. So $Z \cup Y = \mathbb{N}$. Thus

$$Z \cup (\textstyle\bigcap\{Y \supseteq X \mid Y \text{ recursive}\}) = \mathbb{N},$$

and hence $Z \cup X = \mathbb{N}$.

Suppose now that $\bar{x}$ and $\bar{y}$ are standard generators for G_Y, where $X \cup Y = \mathbb{N}$. If $\bar{u}, \bar{v} \in M$ are standard generators for G_Z, for some $Z \supseteq X$, then

$$Y \cup Z \supseteq Y \cup X = \mathbb{N}.$$

Therefore, $St(\bar{u}, \bar{v}; \Delta_1)$ true in M implies that $U(\bar{x}, \bar{y}, \bar{u}, \bar{v})$ is true in M. Thus, (2) is a formula of the form $St(x, y; \Delta_4)$, and (i)$\Rightarrow$(iv).

If a formula $St(x, y; \Delta_4)$ exists, then so does

$$St(x, y; \{Y \mid \mathbb{N}\backslash X \subseteq Y\})$$

and hence, since (i)$\Rightarrow$(iv), so does

$$St\big(x, y; \{Y \mid (\mathbb{N}\backslash X) \cup Y = \mathbb{N}\}\big),$$

which is of the form $St(x, y; \Delta_1)$. So (i)$\Leftrightarrow$(iv). □

We let $\mathfrak{A}^*$ denote the set of all $X \subseteq \mathbb{N}$ for which there exist formulae of the form (i), (ii), (iii), and (iv).

Lemma 9.11 $\mathfrak{A} \subseteq \mathfrak{A}^* \subseteq \bar{\mathfrak{A}}$

Proof If (i) and (ii) exist then

$$St(x, y; \Delta_1) \,\&\, St(x, y; \Delta_2)$$

is a formula of the form $St(x, y; X)$. So $\mathfrak{A}^* \subseteq \bar{\mathfrak{A}}$.

To show that $\mathfrak{A} \subseteq \mathfrak{A}^*$, we show that $\mathfrak{A}^*$ is closed under complementation and recursive enumerability.

By Lemma 9.9, $\emptyset \in \mathfrak{A}^*$, because formulae of the form $St(x, y)$ are also of the form $St(x, y; \{Y \mid \emptyset \subseteq Y\})$.

If a formula of the form $St(x, y; \Delta_1)$ exists for X, then a formula of the form $St(x, y; \Delta_4)$ exists for $\mathbb{N}\backslash X$. So $\mathfrak{A}^*$ is closed under complementation.

For any $X, Y \subseteq \mathbb{N}$, if $Y \leqslant_e X$ then $Y = X\varepsilon_j$, for some $j \in \mathbb{N}$. (Recall that $\varepsilon_j : \mathscr{P}(\mathbb{N}) \to \mathscr{P}(\mathbb{N})$ is given by $X\varepsilon_j = U_j(x)$, see page 128.) First we show that, for any two sets V and W, there is a formula $T(z, w, u, v)$ that is true at $(M, \bar{z}, \bar{w}, \bar{u}, \bar{v})$ if and only if $\bar{u}$ and $\bar{v}$ are standard generators for G_V, with $\bar{z}$ and $\bar{w}$ being standard generators for G_W and with $V\varepsilon_j \subseteq W$. We then show that, for any $W \subseteq \mathbb{N}$, the inclusion $Y \subseteq W$ holds if and only if, for all recursive sets $V \subseteq X$, we have $V\varepsilon_j \subseteq W$. Finally, we use arguments, similar to those in the proof of Lemma 9.10 to show that, if $X \in \mathfrak{A}^*$, then there exists a formula of the form

$$St(z, w; \{W \mid Y \subseteq W\}),$$

and so $Y \in \mathfrak{A}^*$.

Take $W, V \subseteq \mathbb{N}$, and suppose that $(\bar{z}, \bar{w})$ and $(\bar{u}, \bar{v})$ are pairs of standard generators for G_W and G_V, respectively. Let

$$P_i = [w, w_{i+1}] \quad \text{and} \quad q_i = [v, v_{i+1}],$$

so that

$$\bar{p}_i = p_i(\bar{z}, \bar{w}) = 1 \Leftrightarrow i \in W, \qquad \bar{q}_i = q_i(\bar{u}, \bar{v}) = 1 \Leftrightarrow i \in V.$$

Let U_j be the jth recursively enumerable subset of $\mathbb{N} \times \mathscr{P}_f(\mathbb{N})$, in some enumeration, and let $\mathscr{S}(z, w, u, v)$ be the recursively enumerable set

$$\{(q_{i_1} = 1 \,\&\, \cdots \,\&\, q_{i_m} = 1) \Rightarrow (p_n = 1) \mid (n, \{i_1, \ldots, i_m\}) \in U_j\},$$

of positive implications. Then $V\varepsilon_j \subseteq W$ if and only if $\bar{z}, \bar{w}, \bar{u}, \bar{v}$ satisfy $\mathscr{S}(z, w, u, v)$. Since $\mathscr{S}(z, w, u, v)$ is recursively enumerable, it can, by Theorem 9.8, be replaced by a finite set

$$\{r_i(z, w, u, v, x_1, \ldots, x_k) = 1, s_j(z, w, u, v, x_1, \ldots, x_k) \neq 1 \mid 1 \leqslant i \leqslant m, \quad 1 \leqslant j \leqslant l\}$$

of equations and inequalities whose solution in any existentially closed group is equivalent to a solution for the implications. Let $T(z, w, u, v)$ be the formula

$$St(z, w) \,\&\, St(u, v) \,\&\, (\exists x_1, \ldots, x_k) \\ (r_1 = 1 \,\&\, \cdots \,\&\, r_m = 1 \,\&\, s_1 \neq 1 \cdots \,\&\, s_l \neq 1)$$

Then $T(z, w, u, v)$ is true at $(M, \bar{z}, \bar{w}, \bar{u}, \bar{v})$ if and only if $(\bar{z}, \bar{w})$ and $(\bar{u}, \bar{v})$ are pairs of standard generators for G_W and G_V respectively, and $V\varepsilon_j \subseteq W$.

For any $W \subseteq \mathbb{N}$, if $Y \subseteq W$ then clearly for all sets $V \subseteq X$,

$$V\varepsilon_j \subseteq X\varepsilon_j = Y \subseteq W.$$

If $n \in Y$, then there exists a set $\{i_1, \ldots, i_m\} = A \subseteq X$ such that $(n, A) \in U_j$. Then certainly $n \in A\varepsilon_j$, and, since A is finite and therefore recursive, it follows that, if $V\varepsilon_j \subseteq W$ for all recursive sets $V \subseteq X$, then $A\varepsilon_j \subseteq W$. So, $Y \subseteq W$ if and only if $V_j \subseteq W$ for all recursive sets $V \subseteq X$.

Now, for $X \in \mathfrak{A}^*$, consider the formula

$$(\forall u, v)(St(u, v; \{V \mid V \subseteq X\}) \Rightarrow T(z, w, u, v)). \tag{1}$$

If V is a recursive set, then $\mathrm{G}_V \in \mathrm{Sk}\, M$, for any existentially closed group M, so there exist $\bar{u}, \bar{v} \in M$ that are standard generators for G_V. If (i) is true at $(M, \bar{z}, \bar{w})$ then, if $V \subseteq X$, the formula $St(\bar{u}, \bar{v}; \{V \mid V \subseteq X\})$ is true in M. So $T(\bar{z}, \bar{w}, \bar{u}, \bar{v})$ is true in M, and thus $\langle \bar{z}, \bar{w} \rangle \cong \mathrm{G}_W$, for some W such that $V\varepsilon_j \subseteq W$. So if (i) is true at $(M, \bar{z}, \bar{w})$, then $\bar{z}$ and $\bar{w}$ are the standard generators for G_W, for some W satisfying $V\varepsilon_j \subseteq W$ for all recursive sets $V \subseteq X$. So, as above, $Y \subseteq W$. Conversely, if $\bar{z}$ and $\bar{w}$ are standard generators of some G_W with $Y \subseteq W$, then, if $\bar{u}$ and $\bar{v}$ are standard generators of G_V, where $V \subseteq X$, we have $V\varepsilon_j \subseteq Y \subseteq W$. So $T(\bar{z}, \bar{w}, \bar{u}, \bar{v})$ is true in M, and (1) is true at $(M, \bar{z}, \bar{w})$. Thus (1) is a formula of the form $St(z, w; \{W \mid Y \subseteq W\})$, and $Y \in \mathfrak{A}^*$. Hence $\mathfrak{A}^*$ is closed under enumeration reducibility, and $\mathfrak{A} \subseteq \mathfrak{A}^*$. □

Thus we have $\mathfrak{A} \subseteq \bar{\mathfrak{A}}$, which we re-state in the following theorem.

Theorem 9.12 *For any arithmetic set X, there exists a first-order formula that is true at $(M, \bar{x}, \bar{y})$ if and only if $\bar{x}$ and $\bar{y}$ are standard generators for* G_X.

We now wish to turn this into a result about finitely generated groups in general.

Lemma 9.13 *If $K = \langle a_1, \ldots, a_r \rangle$ and $G = \langle b_1, \ldots, b_s \rangle$ are finitely generated groups with* $\mathrm{Rel}\, G \leqslant^* \mathrm{Rel}\, K$, *and if there exists a formula $g(x_1, \ldots, x_r)$ that is true at $(M, \bar{x}_1, \ldots, \bar{x}_r)$, for any existentially closed* *group M, if and only if* $\operatorname{typ}(a_1, \ldots, a_r) = \operatorname{typ}(\bar{x}_1, \ldots, \bar{x}_r)$, *then there exists a formula* $h(y_1, \ldots, y_s)$ *that is true at* $(M, \bar{y}_1, \ldots, \bar{y}_s)$ *if and only if* $K \in \operatorname{Sk} M$ *and* $\operatorname{typ}(b_1, \ldots, b_s) = \operatorname{typ}(\bar{y}_1, \ldots, \bar{y}_s)$.

Proof Let $\mathrm{W}(\boldsymbol{x})$ and $\mathrm{W}(\boldsymbol{y})$ be the sets of all words on the variables $x_1, \ldots, x_r$ and $y_1, \ldots, y_s$, respectively. By definition, there exist recursively enumerable sets

$$U \subseteq \mathrm{W}(\boldsymbol{y}) \times \mathcal{P}_{\mathrm{f}}(\mathrm{W}(\boldsymbol{x})) \quad \text{and} \quad V \subseteq \mathrm{W}(\boldsymbol{y}) \times \mathcal{P}_{\mathrm{f}}(\mathrm{W}(\boldsymbol{x})) \times (\mathrm{W}(\boldsymbol{x}) \cup \{\infty\})$$

such that $u \in \operatorname{Rel}(b_1, \ldots, b_s)$ if and only if, for some $A \subseteq \operatorname{Rel}(a_1, \ldots, a_r)$, we have $(u, A) \in U$, and such that $u \in \operatorname{Nonrel}(b_1, \ldots, b_s)$ if and only if, for some $A \subseteq \operatorname{Rel}(a_1, \ldots, a_r)$ and for some $v \notin \operatorname{Rel}(a_1, \ldots, a_r)$, we have $(u, A, v) \in V$.

Now, the sets

$$\mathcal{S}_1 = \{(w_1(\boldsymbol{x}) = 1 \,\&\, \cdots \,\&\, w_k(\boldsymbol{x}) = 1) \Rightarrow (u(\boldsymbol{y}) = 1) \mid (u, \{w_1, \ldots, w_k\}) \in U\},$$

$$\mathcal{S}_2 = \{(w_1(\boldsymbol{x}) = 1 \,\&\, \cdots \,\&\, w_k(\boldsymbol{x}) = 1 \,\&\, u(\boldsymbol{y}) = 1) \Rightarrow (v(\boldsymbol{x}) = 1) \mid (u, \{w_1, \ldots, w_k\}, v) \in V, \quad v \neq \infty\},$$

$$\mathcal{S}_3 = \{(w_1(\boldsymbol{x})1 = \,\&\, \cdots \,\&\, w_k(\boldsymbol{x}) = 1 \,\&\, u(\boldsymbol{y}) = 1) \Rightarrow \Omega \mid (u, \{w_1, \ldots, w_k\}, \infty) \in V\}$$

are recursively enumerable, and $\mathcal{S}_1 \cup \mathcal{S}_1 \cup \mathcal{S}_3$ is satisfied by the group elements $\bar{x}_1, \ldots, \bar{x}_r, \bar{y}_1, \ldots, \bar{y}_s$ if

$$\operatorname{Rel}(\bar{y}_1, \ldots, \bar{y}_s) \leqslant^* \operatorname{Rel}(\bar{x}_1, \ldots, \bar{x}_r) \quad \text{via } (U, V).$$

By Theorem 9.8, there is a finite set

$$\mathcal{S} = \{p_i(\boldsymbol{x}, \boldsymbol{y}, z_1, \ldots, z_t) = 1 \mid 1 \leqslant i \leqslant n\} \cup \{q_j(\boldsymbol{x}, \boldsymbol{y}, z_1, \ldots, z_t) \mid 1 \leqslant j \leqslant m\}$$

that has a solution in an existentially closed group M if and only if $\mathcal{S}_1 \cup \mathcal{S}_2 \cup \mathcal{S}_3$ has a solution in M. Let

$$P = (\exists z_1, \ldots, z_t)(p_1 = 1 \,\&\, \cdots \,\&\, p_n = 1 \,\&\, q_1 \neq 1 \,\&\, \cdots \,\&\, q_m \neq 1)$$

Let g be the formula that is true at $(M, \bar{x}_1, \ldots, \bar{x}_r)$ if and only if $\operatorname{typ}(a_1, \ldots, a_r) = \operatorname{typ}(\bar{x}_1, \ldots, \bar{x}_r)$, and let

$$h(y_1, \ldots, y_s) = (\exists x_1, \ldots, x_r)(P \,\&\, g).$$

We show that, for any existentially closed group M, the formula h is true at $(M, \bar{y}_1, \ldots, \bar{y}_s)$ is and only if $K \in \operatorname{Sk} M$ and

$$\operatorname{typ}(b_1, \ldots, b_s) = \operatorname{typ}(\bar{y}_1, \ldots, \bar{y}_s).$$

If h is true at $(M, \bar{y}_1, \ldots, \bar{y}_s)$, then, for some $\bar{x}_1, \ldots, \bar{x}_r$ in M, the formula $g(\bar{x}_1, \ldots, \bar{x}_r)$ is true, and so $\operatorname{typ}(a_1, \ldots, a_r) = \operatorname{typ}(\bar{x}_1, \ldots, \bar{x}_r)$ and $K \in \operatorname{Sk} M$. Also, $\operatorname{Rel}(\bar{x}_1, \ldots, \bar{x}_r) = \operatorname{Rel}(a_1, \ldots, a_r)$. Further, $P(\bar{x}_1, \ldots, \bar{x}_r, \bar{y}_1, \ldots, \bar{y}_s)$ is true in M, so $\bar{x}_1, \ldots, \bar{y}_s$ satisfy $\mathcal{S}_1 \cup \mathcal{S}_2 \cup \mathcal{S}_3$. If $u \in \operatorname{Rel}(b_1, \ldots, b_s)$, then there exists $\{w_1, \ldots, w_k\} \subseteq \operatorname{Rel}(a_1, \ldots, a_r)$ such that

$$\big(w_1(\boldsymbol{x}) = 1 \;\&\; \cdots \;\&\; w_k(\boldsymbol{x}) = 1\big) \Rightarrow \big(u(\boldsymbol{y}) = 1\big)$$

belongs to $\mathcal{S}_1$. Since $\bar{x}_1, \ldots, \bar{y}_s$ satisfy $\mathcal{S}_1$, and since

$$\operatorname{Rel}(a_1, \ldots, a_r) = \operatorname{Rel}(\bar{x}_1, \ldots, \bar{x}_r),$$

we have that $u(\bar{y}_1, \ldots, \bar{y}_s) = 1$. So $\operatorname{Rel}(b_1, \ldots, b_s) \subseteq \operatorname{Rel}(\bar{y}_1, \ldots, \bar{y}_s)$. Similarly, from $\mathcal{S}_2$ and $\mathcal{S}_3$, we see that

$$\operatorname{Nonrel}(b_1, \ldots, b_s) \subseteq \operatorname{Nonrel}(\bar{y}_1, \ldots, \bar{y}_s),$$

and thus, since

$$\operatorname{Rel}(b_1, \ldots, b_s) \cup \operatorname{Nonrel}(b_1, \ldots, b_s) = W(\boldsymbol{y}),$$

we have that $\operatorname{Rel}(b_1, \ldots, b_s) = \operatorname{Rel}(\bar{y}_1, \ldots, \bar{y}_s)$. So

$$\operatorname{typ}(b_1, \ldots, b_s) = \operatorname{typ}(\bar{y}_1, \ldots, \bar{y}_s).$$

If $\{\bar{y}_1, \ldots, \bar{y}_s\} \subseteq M$ is such that $\operatorname{typ}(\bar{y}_1, \ldots, \bar{y}_s) = \operatorname{typ}(b_1, \ldots, b_s)$, and, if $K \in \operatorname{Sk} M$, choose $\bar{x}_1, \ldots, \bar{x}_r$ in M such that

$$\operatorname{typ}(a_1, \ldots, a_r) = \operatorname{typ}(\bar{x}_1, \ldots, \bar{x}_r).$$

Then $g(\bar{x}_1, \ldots, \bar{x}_r)$ is true in M, and

$$\operatorname{Rel}(\bar{y}_1, \ldots, \bar{y}_s) = \operatorname{Rel}(b_1, \ldots, b_s) \leqslant^* \operatorname{Rel}(a_1, \ldots, a_r) = \operatorname{Rel}(\bar{x}_1, \ldots, \bar{x}_r).$$

So, $\operatorname{Rel}(\bar{y}_1, \ldots, \bar{y}_s) \leqslant^* \operatorname{Rel}(\bar{x}_1, \ldots, \bar{x}_r)$ via (U, V), and thus $P(\bar{x}_1, \ldots, \bar{x}_r, \bar{y}_1, \ldots, \bar{y}_s)$ is true in M. Hence, h is true at $(M, \bar{y}_1, \ldots, \bar{y}_s)$, and h is the required formula. □

We now use Theorem 9.12 and Lemma 9.13 to prove the result that we have been aiming at.

Theorem 9.14 *For any finitely generated group $G = \langle b_1, \ldots, b_s \rangle$ such that* Rel G *is arithmetic, there exists a first-order sentence*

$$f = (\exists y_1, \ldots, y_s)h$$

such that h is true at $(M, \bar{y}_1, \ldots, \bar{y}_s)$ if and only if

$$\operatorname{typ}(\bar{y}_1, \ldots, \bar{y}_s) = \operatorname{typ}(b_1, \ldots, b_s).$$

So, in particular, f is true in an existentially closed group M if and only if $G \in \operatorname{Sk} M$.

Proof Let

$$K = G_{\text{Rel}\,G} = \langle a_1, a_2 \rangle.$$

Since Rel G is arithmetic, then, by Theorem 9.12, there exists a formula $g(x_1, x_2)$ which is true at $(M, \bar{x}_1, \bar{x}_2)$ if and only if

$$\text{typ}\,(\bar{x}_1, \bar{x}_2) = \text{typ}\,(a_1, a_2).$$

So we can chose, as in Lemma 9.13, a formula $h(y_1, \dots, y_s)$ that is true at $(M, \bar{y}_1, \dots, \bar{y}_s)$ if and only if

$$\text{typ}\,(\bar{y}_1, \dots, \bar{y}_s) = \text{typ}\,(b_1, \dots, b_s)$$

and $K \in \text{Sk}\,M$.

Let f be the sentence $(\exists y_1, \dots, y_s)h$. Then, if f is true in an existentially closed group M, we can find $\bar{y}_1, \dots, \bar{y}_s$ in M such that

$$\langle \bar{y}_1, \dots, \bar{y}_s \rangle \cong \langle b_1, \dots, b_s \rangle = G.$$

Conversely, if $G \in \text{Sk}\,M$, we can find $\bar{y}_1, \dots, \bar{y}_s$ in M such that typ $(\bar{y}_1, \dots, \bar{y}_s) = \text{typ}\,(b_1, \dots, b_s)$ and, since Rel $G \equiv^*$ Rel K, we have $K \in \text{Sk}\,M$. So h is true at $(M, \bar{y}_1, \dots, \bar{y}_s)$. Thus, f is true in M if and only if $G \in \text{Sk}\,M$, as required. □

Example Suppose that $X \subseteq \mathbb{N}$ is an arithmetic set, and let M_X be the unique countable existentially closed group such that $G \in \text{Sk}\,M_X$ if and only if G is finitely generated and Rel $G \leqslant_e X$. From Corollary 6.4 and Theorem 1.12, we see that there exists a subgroup $\langle a, b \rangle$ of M_X such that any group belonging to Sk M_X can be embedded in $\langle a, b \rangle$, i.e. $\text{Sk}\,M_X = \text{Sk}\,\langle a, b \rangle$. Since Rel $(a, b) \leqslant_e X$, we have that Rel (a, b) is arithmetic, and so, from Theorem 9.14, we see that there exists a formula $h(x, y)$ that is true at $(M, \bar{x}, \bar{y})$ if and only if

$$\text{typ}\,(\bar{x}, \bar{y}) = \text{typ}\,(a, b),$$

when M is existentially closed. Recall, from Example 1 of Section 9.3, the formula

$$k(x, y) = (\forall u, v)(\exists s)(\forall t)\big((t^{-1}xt = x \,\&\, t^{-1}yt = y) \Rightarrow (t^{-1}s^{-1}ust = s^{-1}us \,\&\, t^{-1}s^{-1}vst = s^{-1}vs)\big),$$

which is true at $(M, \bar{x}, \bar{y})$ if and only if Sk $\langle \bar{x}, \bar{y} \rangle = \text{Sk}\,M$. Let f be the sentence

$$(\exists x, y)\big(h(x, y) \,\&\, k(x, y)\big).$$

If f is true in an existentially closed group M, then there exist $\bar{x}, \bar{y} \in M$ such that $h(\bar{x}, \bar{y})$ and $k(\bar{x}, \bar{y})$ are true in M. So typ $(\bar{x}, \bar{y}) = \text{typ}\,(a, b)$ and $\text{Sk}\,M_X = \text{Sk}\,\langle a, b \rangle = \text{Sk}\,\langle \bar{x}, \bar{y} \rangle = \text{Sk}\,M$.

Clearly, if $\operatorname{Sk} M = \operatorname{Sk} \mathrm{M}_X$, then $\langle a, b \rangle \in \operatorname{Sk} M$, so there exist $\bar{x}, \bar{y} \in M$ such that $\operatorname{typ}(a, b) = \operatorname{typ}(\bar{x}, \bar{y})$, and

$$\operatorname{Sk} M = \operatorname{Sk} \mathrm{M}_X = \operatorname{Sk} \langle a, b \rangle = \operatorname{Sk} \langle \bar{x}, \bar{y} \rangle.$$

So h and k are true at $(M, \bar{x}, \bar{y})$, and f is true in M.

Thus f is true in M if and only if $\operatorname{Sk} M = \operatorname{Sk} \mathrm{M}_X$, which, if M is countable, is if and only if $M \cong \mathrm{M}_X$. So, given an arithmetic set X, there exists a sentence that is true in a countable existentially closed group M if and only if $M \cong \mathrm{M}_X$.

Exercise (non-trivial) Construct a first-order sentence that is true in a countable existentially closed group M if and only if $M \cong \mathrm{M}_X$, for *some* arithmetic set $X \subseteq \mathbb{N}$.

9.5 GENERIC THEORY

To conclude, we return our attention to those existentially closed groups which can be constructed as the yields of games.

We recall that a game is played with respect to an acceptable code of rules R, i.e. a code of rules with respect to which:

(i) For all properties $\mathscr{P}$ of groups either $\mathscr{P}$ or (not $\mathscr{P}$) is R-enforceable.

(ii) If each of a countable set of properties is R-enforceable, then so is their conjunction.

(iii) Being existentially closed is R-enforceable.

Each sentence of first-order group theory corresponds to a group property. So, given any sentence, we can force it or its negation to be part of the first-order theory of a yield. There are only countably many sentences, so (i) and (ii) imply that we can enforce a whole first-order theory on the yield of a game. By (ii) and (iii), we can also enforce that the yield be existentially closed, so that any theory enforceable under an acceptable code of rules will be the first-order theory of some existentially closed group.

If R is an acceptable code of rules, we say that a *first-order theory is R-generic* if it is R-enforceable, and that a *group is R-generic* if its first-order theory is R-generic. For each R, there is only one R-generic theory, since $\mathscr{P}$ and (not $\mathscr{P}$) cannot both be R-enforceable. But there may be many R-generic groups.

We recall three codes of rules from Chapter 7.

The finite code of rules requires each player to choose only finitely many equations and inequalities at each move.

The $(\varnothing, \varnothing)$ code of rules requires the players to choose equations $\{u = 1 \mid u \in U\}$ and inequalities $\{v \neq 1 \mid v \in V\}$ such that U and V

have finite support, and $(U, V) \leqslant^* (\varnothing, \varnothing)$. From the second example below Lemma 7.1, we see that the finite-generic theory and the $(\varnothing, \varnothing)$-generic theory are the same, and a group is $(\varnothing, \varnothing)$-generic if and only if it is finite-generic.

The stable code of rules requires each player to choose, in turn, a finitely generated group which contains all the groups previously chosen. The yield constructed under these rules is the union of the groups chosen during the game. As a direct corollary of the next lemma, we will see that the sentences that are stably enforceable are exactly those sentences that are stably true. Thus the stable-generic theory is the set of stably true sentences of L.

Before continuing, we wish to digress slightly with the following discussion.

Note We are not using the term 'generic' in exactly the same way as many of our references (Ziegler, Robinson, Macintyre, etc.) do, but since the generic groups have been so important in the study of existentially closed groups, we feel that it is necessary to explain how the traditional generic groups are connected with our R-generic groups.

The following discussion is aimed primarily at the reader who is already familiar with generic groups.

Generic groups were first studied by Robinson (1969, 1970, 1971) using 'forcing' in model theory. It is a consequence of Robinson's definition that generic groups are always existentially closed. Our definition allows an R-generic group not to be existentially closed; in fact, in view of Theorem 9.1, for each R there *must* be an R-generic group which is not existentially closed. We will show that Robinson's generic groups, i.e. those obtained by finite forcing, are exactly the existentially closed $(\varnothing, \varnothing)$-generic groups, Initially we also expected that the infinite-generic groups, i.e. those obtained by infinite forcing, would turn out to be exactly the existentially closed stable-generic groups. But this is not the case. In fact, we will show that, although every infinite-generic group is an existentially closed stable-generic group, there must exist stable-generic groups that are existentially closed but not infinite-generic.

We shall take Zeigler's (1980) definition of generic groups, and we shall use the results that he has proved. The symbol $\Vdash$ will be used for both the finite and infinite forcing relations.

Let θ be any formula in L. We will write $\theta(x_0, \ldots, x_n)$ to indicate that all the free variables in θ lie in the set $\{x_0, \ldots, x_n\}$.

If $(y_0, \ldots, y_n)$ is any sequence of symbols, we write $\theta(y_0, \ldots, y_n)$ (or just $\theta(\mathbf{y})$) for the result of substituting y_i for x_i $(0 \leqslant i \leqslant n)$ in θ. For any group G, any formula $\theta(x_0, \ldots, x_n)$ in L, and any sequence $(a_0, \ldots, a_n)$ of elements in G, we write $G \vDash \theta(a_0, \ldots, a_n)$ when θ is true at $(G, a_0, \ldots, a_n)$.

First we consider the finite forcing case. We will show that the generic groups are exactly the finite-generic existentially closed groups; this then gives the result, since the finite-generic and $(\emptyset, \emptyset)$-generic theories are the same.

Let P be any finite position, and let $\theta(x_0, \dots, x_n)$ be any formula in L. Instead of introducing new constants, we will do our forcing relative to the potential group generators, $g_0, g_1, g_2, \dots$. Following Ziegler, we write $P \Vdash \theta(\boldsymbol{g})$ if, in any game played under the finite rules that begins with a position $Q \supseteq P$, the second player has a strategy ensuring that $\theta(\boldsymbol{g})$ is true in Y, the yield of the game. Equivalently, $P \Vdash \theta(\boldsymbol{g})$ if, in any finite game in which his first move is to the position P, the first player has a strategy that ensures $Y \models \theta(\boldsymbol{g})$.

A group M is generic if, for all formulae $\theta(x_0, \dots, x_n)$ and for all sequences $(a_0, \dots, a_n)$ in M, the assertion $M \models \theta(\boldsymbol{a})$ holds if and only if, for some finite position $P(g_0, \dots, g_m)$, and for some sequence $(b_{n+1}, \dots, b_m)$ in M, we have

$$M \models P(a_0, \dots, a_n, b_{n+1}, \dots, b_m) \quad \text{and} \quad P \Vdash \theta(g_0, \dots, g_n).$$

(This is Ziegler's definition, and he shows (1980) that it is equivalent to Robinson's definition of a generic group.)

A sentence ψ belongs to the finite-generic theory if and only if $\emptyset \models \psi$. If ψ does belong to the finite-generic theory, and if M is a generic group, then we can take $P = \emptyset$, in the definition, to see that $M \models \psi$. If ψ is not in the finite-generic theory, then $\neg\psi$ is, so $M \models \neg\psi$ and $M \not\models \psi$. Thus the first-order theory of a generic group is the finite-generic theory, and so we see that every generic group is $(\emptyset, \emptyset)$-generic. Ziegler (1980, p. 550) shows that every generic group is existentially closed, and also (1980, p. 552) that if G is an existentially closed group with the same first-order theory as some generic group, then G is generic. Thus $(\emptyset, \emptyset)$-generic existentially closed groups are generic.

The infinite forcing case requires more work because we do not have a 'game-theoretic' definition of infinite-generic groups.

Let F be any group. For any sequence $(a_0, \dots, a_n)$ in F, we define $F \Vdash \theta(a_0, \dots, a_n)$, for a formula $\theta(x_0, \dots, x_n)$, inductively as follows:

(a) If θ is quantifier-free, then $F \Vdash \theta(\boldsymbol{a})$ if and only if $F \models \theta(\boldsymbol{a})$.

(b) If $\theta = \psi_1 \wedge \psi_2$, then $F \Vdash \theta(\boldsymbol{a})$ if and only if $F \Vdash \psi_1(\boldsymbol{a})$ and $F \Vdash \psi_2(\boldsymbol{a})$.

(c) If $\theta = \psi_1 \vee \psi_2$, then $F \Vdash \theta(\boldsymbol{a})$ if and only if $F \Vdash \psi_1(\boldsymbol{a})$ or $F \Vdash \psi_2(\boldsymbol{a})$.

(d) If $\theta = \neg\psi$, then $F \Vdash \theta(\boldsymbol{a})$ if and only if, for all groups $G \supseteq F$,

$$G \not\Vdash \psi(\boldsymbol{a}).$$

(e) *If* $\theta = (\exists x_{n+1})\psi(x_0, \dots, x_{n+1})$, then $F \Vdash \theta(\boldsymbol{a})$ if and only if for some $b \in F$,

$$F \Vdash \psi(a_0, \dots, a_n, b).$$

A group M is infinite-generic if, for all formulae $\theta(x_0, \ldots, x_n)$ and all sequences $(a_0, \ldots, a_n)$ in M, the assertion $M \models \theta(\boldsymbol{a})$ holds if and only if $M \Vdash \theta(\boldsymbol{a})$.

Note 1. Parts (b), (c), and (d), above, are consistent. To see this, suppose that

$$F \Vdash \neg(\neg\psi_1(\boldsymbol{a}) \wedge \neg\psi_2(\boldsymbol{a}))$$

then, $F \nVdash \neg\psi_1(\boldsymbol{a}) \wedge \neg\psi_2(\boldsymbol{a})$. It is easily checked that

$$G \supseteq F \text{ and } F \Vdash \theta(\boldsymbol{a}) \quad \Rightarrow \quad G \Vdash \theta(\boldsymbol{a}).$$

Suppose that there exist $G \supseteq F$ and $H \supseteq F$ such that

$$G \Vdash \neg\psi_1(\boldsymbol{a}), \qquad H \Vdash \neg\psi_2(\boldsymbol{a}).$$

Then $K = G *_{G \cap H} H \supseteq F$, and

$$K \Vdash \neg\psi_1(\boldsymbol{a}) \wedge \neg\psi_2(\boldsymbol{a}),$$

which is a contradiction, since $F \Vdash \neg(\neg\psi_1(\boldsymbol{a}) \wedge \neg\psi_2(\boldsymbol{a}))$. So, we may suppose that $G \nVdash \neg\psi_1(\boldsymbol{a})$, for all $G \supseteq F$, and hence,

$$F \Vdash \psi_1(\boldsymbol{a}).$$

2. In general, the quantifier $\forall$ is treated as being $\neg\exists\neg$, so that $F \Vdash (\forall y)\psi$ if and only if $F \Vdash (\neg\exists y)(\neg\psi)$.

When we are dealing with infinite-generic groups we are able to prove the following useful result. If M and N are infinite-generic groups with $M \subseteq N$ then:

$$M \models (\forall y)\psi(\boldsymbol{a}, y) \Rightarrow N \models (\forall y)\psi(\boldsymbol{a}, y). \tag{1}$$

For, $M \models (\neg\exists y)\neg\psi(\boldsymbol{a}, y)$ implies that $M \Vdash (\neg\exists y)\neg\psi(\boldsymbol{a}, y)$ and hence that $N \nVdash (\exists y)\neg\psi(\boldsymbol{a}, y)$. So $N \not\models (\exists y)\neg\psi(\boldsymbol{a}, y)$, giving the result.

Now, let $P = (U, V)$ be any stable-legal position. Let W be the set of all words on the g_i and their inverses, let $(w_0, \ldots, w_n)$ be any sequence of elements of W, and let $\theta(x_0, \ldots, x_n)$ be any formula in L. We write $P \Vdash_\infty \theta(w_0, \ldots, w_n)$ if, in any game played under the stable rules and beginning with some position $P_0 \supseteq P$, the second player has a strategy ensuring that $Y \models \theta(w_0, \ldots, w_n)$, where Y is the yield of the game. We note some properties of $\Vdash_\infty$.

It is clear from the definition that if $Q \supseteq P$, then

$$P \Vdash_\infty \theta(w_0, \ldots, w_n) \Rightarrow Q \Vdash_\infty \theta(w_0, \ldots, w_n). \tag{2}$$

If π is any permutation of the g_i, let $w\pi = w(g_0\pi, g_1\pi, \ldots)$, let $U\pi = \{w\pi \mid w \in U\}$, let $V\pi = \{w\pi \mid w \in V\}$, and let $P\pi = (U\pi, V\pi)$. Then it is easily seen that

$$P \Vdash_\infty \theta(w_0, \ldots, w_n) \Leftrightarrow P\pi \Vdash_\infty \theta(w_0\pi, \ldots, w_n\pi). \tag{3}$$

An obvious modification of the standard argument, which proves that every member of a countable set of properties is enforceable if and only if their intersection is enforceable, gives that

$$P \Vdash_\infty \theta_1 \wedge \theta_2 \Leftrightarrow P \Vdash_\infty \theta_1 \quad \text{and} \quad P \Vdash_\infty \theta_2. \tag{4}$$

And, since W is countable, we also get that

$$P \Vdash_\infty (\forall u)\psi(w_0, \ldots, w_n, u) \Leftrightarrow \text{for all } u \in W,\ P \Vdash_\infty \psi(w_0, \ldots, w_n, u). \tag{5}$$

Clearly,

$$P \Vdash_\infty \neg\theta \Rightarrow P \nVdash_\infty \theta, \tag{6}$$

and, if $P \nVdash_\infty \theta$, then, in a game beginning with some position $Q \supseteq P$, the first player has a strategy that ensures $Y \models \neg\theta(w_0, \ldots, w_n)$. So

$$P \nVdash_\infty \theta \Rightarrow \exists Q \supseteq P,\ Q \Vdash_\infty \neg\theta. \tag{7}$$

From now on, for any group $\langle b_0, \ldots, b_m \rangle$, we will take $\mathrm{Rel}\,(b_0, \ldots, b_m)$ and $\mathrm{Nonrel}\,(b_0, \ldots, b_m)$ to be sets of words on $g_0, \ldots, g_m$, i.e.

$$\mathrm{Rel}\,(b_0, \ldots, b_m) = \{w(g_0, \ldots, g_m) \mid w(\boldsymbol{b}) = 1\},$$

and

$$\mathrm{Nonrel}\,(b_0, \ldots, b_m) = \{w(g_0, \ldots, g_m) \mid w(\boldsymbol{b}) \neq 1\}.$$

We aim to show that, for any formula $\theta(x_0, \ldots, x_n)$, for any infinite-generic group M, and for any sequence $(a_0, \ldots, a_n)$ in M,

$$M \models \theta(\boldsymbol{a}) \quad \text{if and only if} \quad P \Vdash_\infty \theta(g_0, \ldots, g_n), \tag{$*$}$$

where $P = (\mathrm{Rel}\,(a_0, \ldots, a_n), \mathrm{Nonrel}\,(a_0, \ldots, a_n))$. The proof is by induction on the complexity of θ, and there are four cases to consider.

1. Suppose that θ is quantifier-free. Then the truth value of θ at $(G, b_0, \ldots, b_n)$, for any group G, depends only on $\mathrm{Rel}\,(b_0, \ldots, b_n)$. In the yield of any game whose first move contains P, we have

$$\mathrm{Rel}\,(g_0, \ldots, g_n) = \mathrm{Rel}\,(a_0, \ldots, a_n).$$

So, clearly, $M \models \theta(\boldsymbol{a})$ if and only if $P \Vdash_\infty \theta(\boldsymbol{g})$.

2. Suppose that $\theta = \psi_1 \wedge \psi_2$. Then, using (4), we see by induction that $M \models \theta(\boldsymbol{a})$ if and only if $P \Vdash_\infty \theta(\boldsymbol{g})$.

Before considering the last two cases, we prove that, in the case where the support of P is $g_0, \ldots, g_n$, if $P \nVdash_\infty \theta(g_0, \ldots, g_n)$ then $P \Vdash_\infty \neg\theta(g_0, \ldots, g_n)$. (It will be obvious from the proof that, in fact, $P \nVdash_\infty \theta(w_0, \ldots, w_n)$ implies $P \Vdash_\infty \neg\theta(w_0, \ldots, w_n)$, for any sequence $(w_0, \ldots, w_n)$ in which each w_i is a word in $g_0, \ldots, g_n$, and their inverses, only.) Suppose that $P \nVdash_\infty \theta(\boldsymbol{g})$. Then, by (7), there is some stable-legal position $Q \supseteq P$, such that $Q \nVdash_\infty \neg\theta(g_0, \ldots, g_n)$. Let $\boldsymbol{S}$ be a strategy that allows the second player to ensure $Y \models \neg\theta(\boldsymbol{g})$ in any game beginning with $Q_0 \supseteq Q$. If P_0 is any stable position containing P, we show that $\boldsymbol{S}$ can be converted into a strategy that allows the second player to ensure $Y \models \neg\theta(\boldsymbol{g})$ in any game beginning with P_0. Choose a permutation π of the g_i, so that $P_0 \cup Q\pi$ is consistent. Since $P \subseteq P_0 \cap Q$, we may suppose that π fixes $g_0, \ldots, g_n$. There is a game beginning with $Q_0 = (P_0 \cup Q\pi)\pi^{-1} \supseteq Q$, so $\boldsymbol{S}$ produces a move (Q_0, Q_1), say. The second player then chooses the position $P_1 = Q_1\pi$, and makes the move (P_0, P_1). Thus, in the standard way, we continue this procedure to get two parallel games, the second of which is mapped onto the original (first) game by π. If Y_1 is the yield of the second game, the map π from Y_1 to Y extends to an isomorphism. Since $Y_1 \models \neg\theta(g_0, \ldots, g_n)$, we have $Y \models \neg\theta(g_0\pi, \ldots, g_n\pi)$. But π fixes $g_0, \ldots, g_n$, so $Y \models \neg\theta(g_0, \ldots, g_n)$, as required. Thus we obtain $P \Vdash_\infty \neg\theta(g_0, \ldots, g_n)$.

3. Suppose that $\theta = \neg\psi$. Then $M \models \theta(\boldsymbol{a})$ if and only if $M \nvDash \psi(\boldsymbol{a})$, which, by induction, is if and only if $P \nVdash_\infty \psi(\boldsymbol{g})$; by the above remark, this is if and only if $P \Vdash_\infty \theta(g_0, \ldots, g_n)$.

4. Finally, suppose that $\theta(x_0, \ldots, x_n) = (\forall x_{n+1})\psi(x_0, \ldots, x_{n+1})$. If $P \Vdash_\infty \theta(g_0, \ldots, g_n)$, then, by (5), $P \Vdash_\infty \psi(g_0, \ldots, g_{n+1})$. For any $b \in M$, let

$$Q = (\mathrm{Rel}(a_0, \ldots, a_n, b), \ \mathrm{Nonrel}\,(a_0, \ldots, a_n, b)).$$

Then $P \subseteq Q$, and so, by (2), $Q \Vdash_\infty \psi(g_0, \ldots, g_{n+1})$. Thus, by induction, $M \models \psi(a_0, \ldots, a_n, b)$. So $M \models \theta(\boldsymbol{a})$. If $P \nVdash_\infty \theta(\boldsymbol{g})$, then, by (5), there is some $u \in W$ such that

$$P \nVdash_\infty \psi(g_0, \ldots, g_n, u).$$

By (7), we can find $Q \supseteq P$ such that

$$Q \Vdash_\infty \neg\psi(g_0, \ldots, g_n, u),$$

and, by (2), we can extend Q, if necessary, so that, for some group

$$H = \langle a_0, \ldots, a_n, c_{n+1}, \ldots, c_m \rangle,$$

we have

$$Q = (\mathrm{Rel}\,(a_0, \ldots, c_m), \quad \mathrm{Nonrel}\,(a_0, \ldots, c_m))$$

and $u \in Q$. Now, there is a formula ψ' in L such that

$$\psi'(x_0, \ldots, x_m) = \psi(x_0, \ldots, x_n, u(x_0, \ldots, x_m)),$$

and then $Q \Vdash_\infty \neg\psi'(g_0, \ldots, g_m)$. Since $P \subseteq Q$, we can form the free product of M and H, amalgamating $\langle a_0, \ldots, a_n \rangle = A$. It is well known that any group can be embedded in an infinite-generic group, so we can find an infinite-generic group N containing $M *_A H$. By induction,

$$N \models \neg\psi(a_0, \ldots, a_n, c_{n+1}, \ldots, c_m),$$

so

$$N \models \neg\psi'(a_0, \ldots, a_n, u(\boldsymbol{a}, \boldsymbol{c}))$$

and $N \nVdash \theta(\boldsymbol{a})$. Then, by (1), $M \nVdash \theta(\boldsymbol{a})$. This gives the result (*).

A sentence ψ in L belongs to the stable-generic theory if and only if $\varnothing \Vdash_\infty \psi$. So, by (*), if M is an infinite-generic group, and if ψ is a sentence in the stable-generic theory, then $M \models \psi$. If ψ is not in the stable-generic theory, then $\neg\psi$ is, and so $M \models \neg\psi$, and $M \nvDash \psi$. Thus, any infinite-generic group is stable-generic and, as is well known, existentially closed.

By definition, the class of existentially closed stable-generic groups is exactly the elementary class of existentially closed groups whose first-order theory is the stable theory, and we have just seen that this class contains the class of all infinite-generic groups. However, we are grateful to A. Macintyre for pointing out that the class of all infinite-generic groups cannot be an elementary class of existentially closed groups. For, it is well known (see, for example, Simmons, 1972) that the class of all existentially closed groups can be axiomatized by a single sentence ψ of $L_{\omega_1,\omega}$. But, by Macintyre (1975), we see that the class of infinite-generic groups cannot be axiomatized by a single sentence of $L_{\omega_1,\omega}$. If there existed a set of first-order sentences which axiomatized the infinite-generic groups within the class of existentially closed groups, then the conjunction of these sentences and the sentence ψ would be a sentence of $L_{\omega_1,\omega}$, and would axiomatize the class of infinite-generic groups. This is contrary to Macintyre's result, so we see that the class of infinite-generic groups cannot be an elementary class of existentially closed groups. Thus, it must be strictly contained within the class of all existentially closed stable-generic groups.

We end this discussion by pointing out that Ziegler (1980) has given complete characterizations of the infinite-generic and generic groups, and studies the generic groups in great depth. We have used some of Zeigler's ideas in the development of the R-generic groups, but, sadly, a large part of his work in this area remains beyond the scope of this book.

Lemma 9.15 *Let $f(x_1, \ldots, x_s)$ be any first-order formula. In a game played under the stable rules, there is a second-player strategy ensuring that the yield Y of the game has the property: for each sequence $(y_1, \ldots, y_n)$ in Y, the formula f is true at $(Y, y_1, \ldots, y_n)$ if and only if f is stably true at $(y_1, \ldots, y_n)$.*

Proof The proof is by induction on the number of quantifiers in f.

If f is quantifier-free, then the truth value of f at $(G, \bar{x}_1, \ldots, \bar{x}_n)$, for any group G, depends only on $\mathrm{Rel}(\bar{x}_1, \ldots, \bar{x}_n)$. So the result follows immediately.

For any formula $h(z_1, \ldots, z_m)$ and any group G, we will say that G has the property $P(h)$ if, for each sequence $(k_1, \ldots, k_m)$ in G, the formula h

is true at $(G, k_1, \ldots, k_m)$ if and only if h is stably true at $(k_1, \ldots, k_m)$. So we need to show that $P(f)$ is stable-enforceable.

Suppose that $f(\boldsymbol{x}) = (Qz)h(\boldsymbol{x}, z)$, where $Q \in \{\exists, \forall\}$, and that $P(h)$ is stable-enforceable. Now, for any group G and any formula h, it is clear that h is true at $(G, k_1, \ldots, k_m)$ if and only if $\neg h$ is false at $(G, k_1, \ldots, k_m)$, and h is stably true at $(k_1, \ldots, k_m)$ if and only if $\neg h$ is not stably true at $(k_1, \ldots, k_m)$. Thus $P(h)$ holds in G if and only if $P(\neg h)$ holds in G. So, considering $\neg f$ and $\neg h$ if necessary, we may suppose that $f = (\forall z)h$.

Let $\mathbf{S}_1$ be a second-player strategy which forces Y to have $P(h)$. We describe a strategy $\mathbf{S}$ that forces Y to have $P(f)$. The idea is to list the set of all sequences $(y_1, \ldots, y_n)$ of elements of $W = W(g_0, g_1, \ldots)$. Then, on his $(2i)$th move, if he can, the second player acts to ensure that f is false at $(Y, y_1, \ldots, y_n)$, where $(y_1, \ldots, y_n)$ is the ith sequence. We will show that, if the second player cannot act to ensure that f is false at $(Y, y_1, \ldots, y_n)$, then f is stably true at $(y_1, \ldots, y_n)$. We will also show that if Y has $P(h)$, then it follows that f is true at $(Y, y_1, \ldots, y_n)$ if f is stably true at $(y_1, \ldots, y_n)$; and this is enough to give the result.

On his $(2i+1)$th moves ($i \in \mathbb{N}$), the second player acts according to $\mathbf{S}_1$ to insure that Y has $P(h)$. Let $(y_1, \ldots, y_n)$ be the ith n-tuple in W. Suppose that the second player is about to make his $2i$th move, and that he is faced with the position

$$(\operatorname{Rel} H, \operatorname{Nonrel} H).$$

Since $\operatorname{Rel} H$ has finite support, we can find $s \in \mathbb{N}$ such that

$$(\operatorname{Rel} H) \cup (\operatorname{Nonrel} H) \cup \{y_1, \ldots, y_n\}$$

has support $g_0, g_1, \ldots, g_s$. If H has an extension

$$K = \langle g_0, \ldots, g_s, \ldots, g_r \rangle,$$

containing $k_1, \ldots, k_m$, such that $\neg g$ is stably true at $(y_1, \ldots, y_n, k_1, \ldots, k_m)$, then

$$Q_1 = (\operatorname{Rel} H, \operatorname{Nonrel} H) \subseteq (\operatorname{Rel} K, \operatorname{Nonrel} K) = Q_2,$$

and the second player makes the move (Q_1, Q_2). If no such K exists, then the second player makes the trivial move (Q_1, Q_1). We denote this strategy by $\mathbf{S}$.

If $(y_1, \ldots, y_n)$ is a sequence in Y, and if f is stably true at $(y_1, \ldots, y_n)$, then, for all $(u_1, \ldots, u_m)$ in Y, we have that h is stably true at $(y_1, \ldots, y_n, u_1, \ldots, u_m)$. Since $\mathbf{S}$ ensures that Y has $P(h)$, it follows that h is true at $(Y, y_1, \ldots, u_m)$ and thus f is true at $(Y, y_1, \ldots, y_n)$.

If f is not stably true at $(y_1, \ldots, y_n)$, then $\neg f$ is true at $(\mathscr{G}, y_1, \ldots, y_n)$, for any *fg*-complete group $\mathcal{G}$ containing $\langle y_1, \dots, y_n \rangle$. In fact, we can and we will choose $\mathcal{G}$ to contain Y. Then, for some $\bar{z}_1, \dots, \bar{z}_n$ in $\mathcal{G}$, the formula $\neg h$ is true at $(\mathcal{G}, y_1, \dots, y_n, \bar{z}_1, \dots, \bar{z}_m)$. We show that, on his $2i$th move, the second player must have acted to force that f is false at $(Y, y_1, \dots, y_n)$.

Suppose that on his $2i$th move, the second player was faced with the position (Rel H, Nonrel H). For some $s \in \mathbb{N}$,

$$(\text{Rel } H) \cup (\text{Nonrel } H) \cup \{y_1, \dots, y_n\}$$

has support $\{g_0, g_1, \dots, g_s\}$, and

$$H \subseteq \langle g_0, \dots, g_s \rangle \subseteq Y \subseteq \mathcal{G}.$$

Also,

$$H \subseteq \langle g_0, \dots, g_s, \bar{z}_1, \dots, \bar{z}_m \rangle \subseteq \mathcal{G}.$$

Let K_1 be the extension of H generated by $\{g_0, \dots, g_s, g_{s+1}, \dots, g_{s+m}\}$ and constructed so that

$$\text{typ}\,(g_0, \dots, g_{s+m}) = \text{typ}\,(g_0, \dots, g_s, \bar{z}_1, \dots, \bar{z}_m).$$

In particular, then,

$$\text{typ}\,(y_1, \dots, y_n, g_{s+1}, \dots, g_{s+m}) = \text{typ}\,(y_1, \dots, y_n, \bar{z}_1, \dots, \bar{z}_m),$$

and so, since $\neg h$ is stably true at $(y_1, \dots, y_n, \bar{z}_1, \dots, \bar{z}_m)$, we see that $\neg h$ is stably true at $(y_1, \dots, y_n, g_{s+1}, \dots, g_{s+m})$. Thus, a group of the type required by **S** exists, and so, on his $2i$th move, the player moves to the position (Rel K, Nonrel K), where $H \subseteq K$ and, for some $\{k_1, \dots, k_m\} \subseteq K$, the formula $\neg h$ is stably true at $(y_1, \dots, y_n, k_1, \dots, k_m)$. Since $K \subseteq Y$ and Y has $P(h)$, it follows that $\neg h$ is true at $(Y, y_1, \dots, y_n, k_1, \dots, k_n)$, and so $\neg f$ is true at $(Y, y_1, \dots, y_n)$. This proves the lemma. □

Corollary 9.16 *A sentence is stable-enforceable if and only if it is stably true.*

Proof If a sentence f is stable-enforceable we can force that f is true in Y and also, by Lemma 9.15, that f is true in Y if and only if f is stably true. Thus, f is stably true.

If f is stably true, then, by Lemma 9.15, we can force that f is true in the yield of a game played under the stable rules. □

The next theorem shows that, for the two codes of rules that we have been considering, there is a wealth of countable generic existentially closed groups—indeed, as many as there could be.

Theorem 9.17 (*i*) *There are* $2^{\aleph_0}$ *countable* $(\varnothing, \varnothing)$*-generic existentially closed groups, no two of which are isomorphic.*

(*ii*) *There are* $2^{\aleph_0}$ *countable stable-generic existentially closed groups, no two of which are isomorphic.*

Proof (i) From the proof of Theorem 7.6, we see that there are $2^{\aleph_0}$ distinct countable existentially closed groups which are the yields of games played under the finite code of rules. Each of these groups can be forced to be finite-generic, and thus $(\varnothing, \varnothing)$-generic.

(ii) There are $2^{\aleph_0}$ finitely generated groups, and each of these can be forced into the yield of some game played under the stable rules. If there were less than $2^{\aleph_0}$ stable-generic countable existentially closed groups, then these groups, between them, would contain less than $2^{\aleph_0}$ finitely generated subgroups, since each countable group contains only countably many finitely generated subgroups. So, we could find a finitely generated group G that does not belong to the skeleton of any stable-generic countable existentially closed group. Since there is a second-player strategy that forces the yield of a game under the stable rules to be existentially closed and stable-generic, and to have G in its skeleton, this would be a contradiction. So we must have $2^{\aleph_0}$ distinct groups of the type required. □

Theorem 9.18 *Let G be a finitely generated group such that* Rel G *is arithmetic, and let M be an existentially closed group.*

(i) *If M is stable-generic, then $G \in$* Sk *M.*

(ii) *If M is $(\varnothing,\varnothing)$-generic, then $G \in$* Sk *M if and only if G has solvable word problem.*

Proof By Theorem 9.14, there exists a first-order sentence f that is true in an existentially closed group M if and only if $G \in \mathrm{Sk}\, M$. Under any code of rules R, either f or $\neg f$ is R-enforceable, so f or $\neg f$ belongs to the R-generic theory. Thus, G belongs to the skeleton of an R-generic existentially closed group if and only if f belongs to the R-generic theory. Thus, to prove (i), we only need to note that we can stable-enforce any finitely generated group to belong to the yield of a game played under the stable rules. So G belongs to the skeleton of every stable-generic existentially closed group.

For (ii), if M is $(\varnothing, \varnothing)$-generic and $G \in \mathrm{Sk}\, M$, then, by Theorem 9.14, G belongs to the skeletons of all $(\varnothing,\varnothing)$-generic existentially closed groups. As in the proof of Theorem 7.6, there exist such groups whose skeletons intersect only in groups with solvable word problem. So G must have solvable word problem. □

It would be nice if codes of rules were somehow characterized by the $\forall_n$-sentences or the $\exists_n$-sentences that are true in the corresponding generic first-order theories. We have no evidence to suggest that this is true but the following theorem may be of some interest in this context.

Theorem 9.19

(i) *A $\forall_3$-sentence is true in a stable-generic group if and only if it is true in all existentially closed groups.*

(ii) *A $\forall_3$-sentence is true in a $(\varnothing, \varnothing)$-generic group if and only if it is true in some existentially closed group.*

Proof In view of Corollary 9.16, (i) is just a re-statement of Theorem 9.5.

To prove (ii), suppose that $(\forall \boldsymbol{x})(\exists \boldsymbol{y})(\forall \boldsymbol{z})f$ is a $\forall_3$-sentence which is true in some existentially closed group M. We will show that we can enforce that the sentence is true in the yield of a game played under the $(\varnothing, \varnothing)$-generic rules, and hence that it belongs to the $(\varnothing, \varnothing)$-generic theory. The converse follows trivially since there exists a $(\varnothing, \varnothing)$-generic existentially closed group.

First we show that, for any sequence $(\bar{x}_1, \ldots, \bar{x}_n)$ in M, there exists a sequence $(\bar{y}_1, \ldots, \bar{y}_m)$ in M such that $(\forall \boldsymbol{z})f$ is stably true at $(\bar{x}_1, \ldots, \bar{x}_n, \bar{y}_1, \ldots, \bar{y}_m)$. We then enumerate all the sequences of length n in $W(g_0, g_1, \ldots)$, and describe how, using Lemma 9.4, on his ith move the second player can act to ensure that $(\forall \boldsymbol{z})f$ is stably true at $(w_1, \ldots, w_n, g_{r+1}, \ldots, g_{r+m})$, where $(w_1, \ldots, w_n)$ is the ith sequence. Since we can find an fg-complete group that contains the yield of the game, we easily see that the result then follows.

For any sequence $(\bar{x}_1, \ldots, \bar{x}_n)$ in M we can find a sequence $(\bar{y}_1, \ldots, \bar{y}_m)$ such that $(\forall \boldsymbol{z})f$ is true at $(M, \bar{x}_1, \ldots, \bar{x}_n, \bar{y}_1, \ldots, \bar{y}_m)$. We write f in the form $\neg P_1 \,\&\, \neg P_2 \,\&\, \cdots \,\&\, \neg P_k$, where each P_i is the conjunction of equations and inequalities involving elements of $W(\boldsymbol{x}, \boldsymbol{y}, \boldsymbol{z})$. Let $\mathcal{G} \supseteq M$ be an fg-complete group. Then, if, for some $\{h_1, \ldots, h_t\} \subseteq \mathcal{G}$, the formula f is false at $(\mathcal{G}, \bar{x}_1, \ldots, \bar{y}_m, h_1, \ldots, h_t)$, we may suppose that

$$P_1 = (u_1 = 1) \,\&\, \cdots \,\&\, (u_\alpha = 1) \,\&\, (v_1 \neq 1) \,\&\, \cdots \,\&\, (v_\beta \neq 1)$$

is true at $(\mathcal{G}, \bar{x}_1, \ldots, h_t)$. We let

$$\bar{u}_i = u_i(\bar{x}_1, \ldots, \bar{x}_n, \bar{y}_1, \ldots, \bar{y}_m, z_1, \ldots, z_t),$$

$$\bar{v}_j = v_j(\bar{x}_1, \ldots, \bar{x}_n, \bar{y}_1, \ldots, \bar{y}_m, z_1, \ldots, z_t).$$

Then the set $\mathcal{S}_1 = \{\bar{u}_i = 1, \quad \bar{v}_j \neq 1 \mid 1 \leqslant i \leqslant \alpha, \quad 1 \leqslant j \leqslant \beta\}$, defined over M and with variables $z_1, \ldots, z_t$, has a solution $h_1, \ldots, h_t$ in $\mathcal{G} \supseteq M$. Thus there is a solution $\bar{z}_1, \ldots, \bar{z}_t$ in M, and P_1 is true at $(M, \bar{x}_1, \ldots, \bar{y}_m, \bar{z}_1, \ldots, \bar{z}_t)$. So $(\forall \boldsymbol{z})f$ is false at $(M, \bar{x}_1, \ldots, \bar{y}_m)$, which is a contradiction. Thus $(\forall \boldsymbol{z})f$ is true at $(\mathcal{G}, \bar{x}_1, \ldots, \bar{y}_m)$ and $(\forall \boldsymbol{z})f$ is stably true at $(\bar{x}_1, \ldots, \bar{x}_n, \bar{y}_1, \ldots, \bar{y}_m)$.

Let $(w_1(\boldsymbol{g}), \ldots, w_n(\boldsymbol{g}))$ be the ith sequence of W, and suppose that, at the beginning of his ith move, the second player is faced with the position (X, Z). We also suppose that $X \cup Z \cup \{w_1, \ldots, w_n\}$ has support

$g_0, g_1, \dots, g_r$. Let

$$\mathcal{S} = \{w(\mathbf{g}) = 1 \mid w \in X\} \cup \{w(\mathbf{g}) \neq 1 \mid w \in Z\}.$$

Then $\mathcal{S}$ is soluble in the yield Y, and hence in $M * Y$ and thus in M. Let $\{\bar{g}_0, \dots, \bar{g}_r\}$ be a set of solutions for $\mathcal{S}$ in M, and let $\bar{x}_i = w_i(\bar{g}_0, \dots, \bar{g}_r)$. As above, there exists a set $(\bar{y}_1, \dots, \bar{y}_m)$ in M, such that $(\forall \mathbf{z})f$ is stably true at $(\bar{x}_1, \dots, \bar{x}_n, \bar{y}_1, \dots, \bar{y}_m)$. By Lemma 9.4, we can find finite subsets

$$A \subseteq \mathrm{Rel}\,(\bar{x}_1, \dots, \bar{y}_m), \qquad B \subseteq \mathrm{Nonrel}\,(\bar{x}_1, \dots, \bar{y}_m),$$

such that $(\forall \mathbf{z})f$ is stably true at $(a_1, \dots, a_n, b_1, \dots, b_m)$ if $A \subseteq \mathrm{Rel}\,(a_1, \dots, b_m)$ and $B \subseteq \mathrm{Nonrel}\,(a_1, \dots, b_m)$, for any group $\langle a_1, \dots, a_n, b_1, \dots, b_m \rangle$.

Let

$$X_A = \{w(w_1, \dots, w_n, g_{r+1}, \dots, g_{r+m}) \mid w \in A\},$$
$$Z_B = \{w(w_1, \dots, w_n, g_{r+1}, \dots, g_{r+m}) \mid w \in B\}.$$

There is a homomorphism $\theta : F\langle g_0, \dots, g_{m+r} \rangle \to M$, given by

$$g_i\theta = \bar{g}_i \quad (0 \leqslant i \leqslant r), \qquad g_{j+r}\theta = \bar{y}_j \quad (1 \leqslant j \leqslant m),$$

where $F = F\langle g_0, \dots, g_{m+r} \rangle$ is a free group of rank $r + m + 1$. Since $\bar{g}_0, \dots, \bar{g}_r$ are solutions of $\mathcal{S}$, and since $A \subseteq \mathrm{Rel}\,(\bar{x}_1, \dots, \bar{y}_m)$ and $B \subseteq \mathrm{Nonrel}\,(\bar{x}_1, \dots, \bar{y}_m)$, it follows that $(X \cup X_A)\theta = \{1\}$ and $v\theta \neq 1$, for all $v \in Z \cup Z_B$. Thus $\langle X \cup X_A \rangle^F \cap \langle Z \cup Z_B \rangle = \varnothing$, and so

$$(X \cup X_A, \quad Z \cup Z_B) = P$$

is consistent. Since X_A and Z_B are finite, and since $(X, Z) \leqslant^* (\varnothing, \varnothing)$, we have $P \leqslant^* (\varnothing, \varnothing)$, so that the second player can, and does, move to the position P. So, for the subgroup $\langle w_1, \dots, w_n, g_{r+1}, \dots, g_{r+m} \rangle$ of Y, since X_A belongs to the relations of Y, and Z_B belongs to the nonrelations of Y, we have that

$$A \subseteq \mathrm{Rel}\,(w_1, \dots, g_{r+m}), \qquad B \subseteq \mathrm{Nonrel}\,(w_1, \dots, g_{r,m}).$$

Thus $(\forall \mathbf{z})f$ is stably true at $(w_1, \dots, w_n, g_{r+1}, \dots, g_{r+m})$.

Let Y be the yield of a game, played under the $(\varnothing, \varnothing)$ rules, in which the second player has played according to the above strategy. For any sequence $(w_1, \dots, w_n)$ in Y, we can find an integer r and an *fg*-complete group $\mathcal{G} \supseteq Y$, such that $(\forall \mathbf{z})f$ is true at $(\mathcal{G}, w_1, \dots, w_n, g_{r+1}, \dots, g_{r+m})$. Thus $(\forall \mathbf{z})f$ is true at $(Y, w_1, \dots, w_n, g_{r+1}, \dots, g_{r+m})$, since f is quantifier-free and since $Y \subseteq \mathcal{G}$. So $(\exists \mathbf{y})(\forall \mathbf{z})f$ is true at $(Y, w_1, \dots, w_n)$, for all $(w_1, \dots, w_n)$ in Y, and thus $(\forall \mathbf{x})(\exists \mathbf{y})(\forall \mathbf{z})f$ is $(\varnothing, \varnothing)$-enforceable, as required. □

Since the $\exists_3$-sentences are precisely the negations of the $\forall_3$-sentences, we obtain directly, from Theorem 9.19, the following result.

Corollary 9.20 (i) *A $\exists_3$-sentence is true in a stable-generic group if and only if it is true in some existentially closed group.*

(ii) *A $\exists_3$-sentence is true in a $(\emptyset, \emptyset)$-generic group if and only if it is true in all existentially closed groups.*

Corollary 9.21 *None of the groups* M_X *is $(\emptyset, \emptyset)$-generic.*

Proof Let $G \in \mathrm{Sk}\,\mathrm{M}_\emptyset \subseteq \mathrm{Sk}\,\mathrm{M}_X$ be a group with unsolvable word problem. Since $G \in \mathrm{Sk}\,\mathrm{M}_\emptyset$, we can embed G in a finitely presented group. So, as in Example 2 of Section 9.3, we can find a $\exists_3$-sentence f (say) that is true in an existentially closed group M if and only if $G \in \mathrm{Sk}\,M$. Since G has unsolvable word problem, there is an existentially closed group in which f is not true. So, by Corollary 9.20(ii), f does not belong to the $(\emptyset, \emptyset)$-generic theory. But $G \in \mathrm{Sk}\,\mathrm{M}_X$, so f belongs to the theory of M_X, and M_X cannot be $(\emptyset, \emptyset)$-generic. □

Note Recall, from Corollary 9.6, that $\forall_2$-sentences and $\exists_2$-sentences have the same truth value in all existentially closed groups. Also, a sentence is true in an R-generic group if and only if it belongs to the R-generic theory. Let $\mathrm{Th}_{\forall n}\, R$ and $\mathrm{Th}_{\exists_n}\, R$ denote, respectively, the sets of all $\forall_n$-sentences and all $\exists_n$-sentences that belong to the R-generic theory. Since there is an R-generic existentially closed group for every acceptable code of rules R, we see from above that we have

$$\mathrm{Th}_{\forall_2}\, R = \mathrm{Th}_{\forall_2}\,(\emptyset, \emptyset), \qquad \mathrm{Th}_{\exists_2}\, R = \mathrm{Th}_{\exists_2}\,(\emptyset, \emptyset),$$

for all acceptable R. So, no two R-generic theories can be separated by a $\exists_2$-sentence or a $\forall_2$-sentence. Also, from Theorem 9.19 and Corollary 9.20, we see that, for all acceptable rules R,

$$\mathrm{Th}_{\forall_3}\,(\text{stable}) \subseteq \mathrm{Th}_{\forall_3}\, R \subseteq \mathrm{Th}_{\forall_3}(\emptyset, \emptyset),$$
$$\mathrm{Th}_{\exists_3}(\emptyset, \emptyset) \subseteq \mathrm{Th}_{\exists_3}\, R \subseteq \mathrm{Th}_{\exists_3}\,(\text{stable}).$$

Example 2 of Section 9.3 and its negation are, respectively, a $\exists_3$-sentence and a $\forall_3$-sentence which are true in some, but not all, existentially closed groups. So, again from Theorem 9.19 and Corollary 9.20, we see that

$$\mathrm{Th}_{\forall_3}\,(\text{stable}) \neq \mathrm{Th}_{\forall_3}\,(\emptyset, \emptyset), \qquad \mathrm{Th}_{\exists_3}(\text{stable}) \neq \mathrm{Th}_{\exists_3}\,(\emptyset, \emptyset).$$

Thus we can separate the stable-generic and $(\emptyset, \emptyset)$-generic theories by looking at the $\forall_3$-sentences belonging to them, or by looking at the $\exists_3$-sentences belonging to them.

Bibliography

1. J. Barwise and A. Robinson, *Completing Theories by Forcing,* Ann. Math. Logic, **2** (1970), 119–142.
2. O. V. Belegradek, *Definability in Algebraically Closed Groups,* Math. Notes of the Academy of Sciences of the U.S.S.R., **16** No. 3 (1974), 813–816.
3. ——, *On Algebraically Closed Groups,* Algebra i Logika, **13** No. 3 (1974), 239–255. [Russian].
4. ——, *Elementary Properties of Algebraically Closed Groups,* Fund. Math. **XCVIII** (1978), 83–101. [Russian].
5. J. L. Britton, *The Word Problem,* Ann. Math., **77** (1963).
6. H. S. M. Coxeter and W. O. J. Moser, *Generators and Relations for Discrete Groups,* Ergebrusse der Mathematik, **14,** *Springer,* (1972).
7. K. Hickin and A. Macintyre, *Algebraically Closed Groups: Embeddings and Centralizers,* in *Word Problems II, The Oxford Book,* ed. S. I. Adian, W. W. Boone, and G. Higman, *North Holland,* (1980), 141–155.
8. K. Hickin, *A.c. Groups: Extensions, Maximal Subgroups, and Automorphisms,* Trans. Am. Math. Soc., **290,** No. 2 (1985), 457–481.
9. G. Higman, B. H. Neumann, and Hanna Neumann, *Embedding Theorems for Groups,* J. London Math. Soc., **24** (1949), 247–254.
10. G. Higman, *Subgroups of Finitely Presented Groups,* Proc. Royal Soc. London (A), **262** (1961), 455–475.
11. J. Hirschfeld and W. H. Wheeler, *Forcing, Arithmetic, Division Rings,* Springer Lecture Notes in Mathematics, *Springer-Verlag,* **454** (1975).
12. R. C. Lyndon and P. E. Schupp, *Combinatorial Group Theory,* Ebgenisse der Mathematik, **89,** *Springer-Verlag,* (1977).
13. A. Macintyre, *On Algebraically Closed Groups,* Ann. Math., **96** (1972), 53–97.
14. ——, *Omitting Quantifier-Free Types in Generic Structures,* J. Sym. Logic, **37** (1972), 512–520.
15. ——, *Martins Axiom Applied to Existentially Closed Groups,* Mathematica Scandinavica, **32** (1973), 46–56.
16. ——, *A Note on Axioms for Infinite-Generic Structures,* J. London Math. Soc., **9** (1974–5), 581–584.
17. ——, *Existentially Closed Structures and Jensen's Principle* $\diamondsuit$, Israel J. Math. **25** (1976), 202–210.
18. C. F. Miller III, *The Word Problem in Quotients of a Group,* unpublished.
19. ——, *Decision Problems in Algebraic Classes of Groups* (*A Survey*), in *Word Problems,* ed. W. W. Boone, F. B. Cannonito and R. C. Lyndon, *North Holland,* (1973).
20. B. H. Neumann, *A Note on Algebraically Closed Groups,* J. London Math. Soc., **27** (1952), 247–249.
21. ——, *The Isomorphism Problem for Algebraically Closed Groups,* in *Word*

Problems, ed. W. W. Boone, F. B. Cannonito and R. C. Lyndon, *North Holland,* (1973).

22. P. M. Neumann, *The SQ-Universality of Some Finitely Presented Groups,* J. Aust. Math. Soc., **XVI** (1973).
23. A. Robinson, *Forcing in Model Theory,* Symposia Mathematica, **5,** *Academic Press,* (1971), 69–82.
24. ——, *Infinite Forcing in Model Theory,* Proc. Second Scandinavian Logic Symposium Oslo 1970, *North Holland,* (1971), 317–340.
25. ——, *Forcing in Model Theory,* Proc. International Congress of Math. In Nice 1970, *Gauthier-Villars,* (1971).
26. ——, *On the Notion of Algebraic Closedness for Noncommutative Groups and Fields,* J. Symbolic Logic, **36** (1971), 441–444.
27. H. Rogers, *Theory Recursive Functions and Effective Computability, McGraw-Hill,* (1967).
28. W. R. Scott, *Algebraically Closed Groups,* Proc. Am. Math. Soc., **2** (1951).
29. S. Shelah and M. Ziegler, *Algebraically Closed Groups of Large Cardinality,* J. Symbolic Logic, **44** (1979).
30. H. Simmons, *Existentially Closed Structures,* J. Symbolic Logic, **37** (1972), 293–310.
31. ——, *The Word Problem for Absolute Presentations,* J. London Math. Soc., **6** (1973), 275–280.
32. ——, *Large and Small Existentially Closed Structures,* J. Symbolic Logic, **41** (1976), 379–390.
33. M. Y. Trofimov, *Definability in Algebraically Closed Systems,* Algebra i Logika, **14** (1975), 320–327. [Russian].
34. C. Wood, *Forcing for Infinitary Languages,* Zeitschrift f. Math. Logic, **18** (1972), 385–402.
35. M. Ziegler, *Algebraisch Abgeschlossene Gruppen,* in *Word Problems II, The Oxford Book,* ed. S. I. Adian, W. W. Boone, and G. Higman, *North Holland,* (1980), 449–576.

Index